T0303284

THE MEASLES
BOOK

THE MEASLES
BOOK

Thirty-Five Secrets the Government and the Media Aren't Telling You about Measles and the Measles Vaccine

Foreword by Robert F. Kennedy Jr.

Skyhorse Publishing

Skyhorse Publishing books may be purchased in bulk at special discounts for sales promotion, corporate gifts, fund-raising, or educational purposes. Special editions can also be created to specifications. For details, contact the Special Sales Department, Skyhorse Publishing, 307 West 36th Street, 11th Floor, New York, NY 10018 or info@skyhorsepublishing.com.

Skyhorse® and Skyhorse Publishing® are registered trademarks of Skyhorse Publishing, Inc.®, a Delaware corporation.

Visit our website at www.skyhorsepublishing.com.

10 9 8 7 6 5 4 3 2 1

Library of Congress Cataloging-in-Publication Data is available on file.

Print ISBN: 978-1-5107-6824-6
Ebook ISBN: 978-1-5107-6825-3

Printed in the United States of America

MESSAGE TO THE READER

This book reveals thirty-five secrets about measles and the measles vaccine. You will be shocked about what you have not been told.

Have the benefits of the measles vaccine been exaggerated? Have the dangers been brushed aside? You will soon know.

Whether you vaccinate or not is a decision that should be made by parents and doctors. Not politicians. Not pharmaceutical companies. After you read this book, you will understand why.

All of these secrets are available with a little digging. The media could report on it today if they wanted. Politicians could discuss it. The pharmaceutical industry could tell the truth. But they don't. They keep it hidden.

That's the purpose of this book—to uncover the facts that have been hidden. Why? Because it affects your child's health and life.

Contents

Preface   ix

Foreword by Robert F. Kennedy Jr.   xiii

Introduction   xv

PART 1: There Is No Need to Panic about Measles   1

PART 2: The Measles Vaccines Are Worth Billions of Dollars
            to Vaccine Makers and the Government   27

PART 3: The Measles Vaccine Can Be Dangerous to Some Children   53

PART 4: The Shocking Ingredients in the Measles Vaccine   123

PART 5: The Autism-Vaccine Secrets You Have Never Been Told   139

PART 6: Vaccine Companies Are Not Held Accountable When
            a Child Is Hurt or Killed by a Vaccine   165

PART 7: The Government Is Forcing a One-Size-Fits-All
            Healthcare Plan That Does Not Fit All   175

Conclusion: You Should Not Be Forced to Vaccinate   201

Glossary   205

Preface

COVID! The plague of the modern era. That's what we are told. The media screams at us day and night about deaths and case counts. The more TV we consume, the more we are whipped up into a frenzy of fear. Fear of others. Fear of the outside. Fear of the air.

But there is hope on the horizon. We are told that poorly tested, unproven, and experimental "vaccines" will save us. They say that once we allow these chemicals and nano-particles to enter our body we are good to go. We are even told that it's safe to allow children, babies, and pregnant women to get the experimental and unproven shots despite the lack of any long-term safety tests.

A deadly contagious disease that only the "vaccine" can save us from. Is this a new story? Is COVID a unique tale?

Unfortunately, it's not.

The COVID story is a remix of an old vaccine song that's been played for over fifty years. It's been amped up beyond anything we have ever seen before, but the lyrics remain the same. They go something like this:

Verse 1: Contagious diseases in the United States are our most dangerous menace and our worst nightmare.

Verse 2: There are no successful ways to prevent or treat these diseases and that's why vaccines are our only hope.

Verse 3: Vaccines are completely safe and effective.

Verse 4: The only people who are against vaccines are wacky nut jobs.

Verse 5: Vaccines contain only healthy ingredients like water.

Chorus: Vaccines are produced by altruistic businesses who are focused on safety and who only care about our health.

Those are pleasant lyrics, but, unfortunately, they are completely false. They were false when the song started playing fifty years ago and they are false today with the COVID-19 vaccines.

Here are the facts:

Verse 1: Contagious diseases in the United States are our most dangerous menace and our worst nightmare.
False. Since the improvements in hygiene and sanitation a century ago, contagious diseases have steadily dropped without vaccines. On the other hand, one million people died from medical errors and drug interactions in the last four years. Where's the news coverage about that? There is none. Perhaps because there are no profits to be had.

Verse 2: There are no successful ways to prevent or treat these contagious diseases and that's why vaccines are our only hope.
False. There are treatments available for every contagious disease. For example, did you know that Vitamin A is the treatment for measles and there are many drugs that prevent and treat COVID, such as Remdesivir?

Verse 3: Vaccines are completely safe and effective.
False. There are literally hundreds of thousands of deaths and injuries related to vaccines. The vaccine death and injury cases are listed in the Vaccine Adverse Event Reporting System (VAERS) government database and are accessible from your computer. As of this writing, there has been a total of 545,338 reports of adverse events from all age groups following COVID vaccines, including 12,366 deaths and 70,105 serious injuries between Dec. 14, 2020 and July 30, 2021. Those are from the COVID vaccine. There are more reports for the measles vaccine.

Verse 4: The only people who are against vaccines are wacky nut jobs.
False. The people who are concerned about vaccine safety are those who have looked at the vaccine injury data, read the clinical trial studies, or have someone in their family injured or killed by a vaccine.

Verse 5: Vaccines contain only healthy ingredients like water.
Not even close. Vaccines contain genetically engineered substances, viruses that have been exposed to hundreds of animals, industrial chemicals, heavy metals, and other toxins. In fact, the COVID vaccines contain MRNA, the genetic material that programs our cells.

Chorus: Vaccines are produced by altruistic businesses who are focused on safety and who only care about our health.

False. Some of the pharmaceutical companies that make vaccines have been found guilty of price fixing, over billing, and data falsification. They have also been sued for wrongful death and have had criminal charges filed against them. And by the way, they are totally indemnified from liability for vaccine injuries. This means that when their vaccines injure or kill someone, they pay nothing.

Once you read this book, you will recognize how the COVID narrative ("the song") is a much amplified version of what's been played again and again in the past with other vaccine campaigns. The song always has one goal: to get us to roll up our sleeve and get injected.

The media hype, the fear mongering, the lack of objectivity, the manipulating of statistics, the ignoring of effective therapies, the censorship, and the propaganda have all been used before in other vaccine campaigns, like measles. The same vaccine song, the same chorus, the same lies. And now, with COVID, it's deafening.

What's past is prologue
—William Shakespeare

Foreword

by Robert F. Kennedy Jr.

Children's Health Defense's *The Measles Book: Thirty-Five Secrets the Government and the Media Aren't Telling You about Measles and the Measles Vaccine* provides the reader with vital, clear information they should have been told long ago.

The readers—American and global consumers of measles vaccines—will learn that they have been misled by the pharmaceutical industry and their captured government agency allies into believing that measles is a deadly disease and that measles vaccines are necessary, safe, and effective.

The Measles Book explodes this propaganda. In concise, factual detail, CHD rips the cloak off Pharma industry slogans to reveal a disease that is rarely life threatening and a vaccine that is largely unnecessary but which carries real risks for the children mandated to take them. Parents are simply not informed of the "real deal" with measles vaccines. The information in *The Measles Book* may come as a shock to many who previously trusted the "public health experts."

Why have the benefits of the measles vaccine been exaggerated and the risks understated? Because these vaccines are a cash cow for an industry that long ago left behind the legacy of true public health pioneers like Dr. Jonas Salk. Big Pharma has devolved into nothing more than a government-enabled cartel. CHD informs the reader that measles vaccines carry risks—but only for the vaccinee and his or her family. Big Pharma is free of virtually all legal liability for damages from the vaccines' harms.

The Measles Book enlightens the reader that mainstream media, Pharmedia, is complicit in protecting Big Pharma and aiding and abetting its fear mongering and racketeering. True investigative journalists should have long ago exposed the facts that CHD is making manifest: Measles outbreaks have been fabricated to create fear that in turn forces government officials to "do something." They then inflict unnecessary and risky vaccines on millions of children for the sole purpose of fattening industry profits. When a child is injured—and many are, but most injuries are not reported—the government and Pharma walk away, denying responsibility and liability for damages. If the child is one of the rare few to obtain compensation from the government for his or her injuries, the reader will be shocked to learn that taxpayers, not Pharma, foot the bill.

Pharma walks away, scot-free.

Grounded in powerful, well-documented evidence, *The Measles Book* compels readers to do what public health cognoscenti fear most—think for themselves.

Introduction

Measles! We all have seen or heard the scary stories about "outbreaks" in the media. It has even been declared a "public health emergency" at various times. Is it true? Are we and our children at risk?

This book will help you answer these questions. You will find out if this is just another example of the media, government, and industry misleading us or whether we really have a lot to worry about.

This book presents reliable medical information from the most credible sources available. It is intended to help you make an informed choice about vaccinating your child. The book focuses on measles, but many of the issues are relevant to other childhood vaccines.

Secrets

The pages you are about to read reveal secrets. Hidden pieces of information that you should know.

What's wrong with keeping this information secret?

Everything! This is the information that parents like us need in order to make informed decisions for our children. We want to make good decisions for our children. We want to know the truth about the risks and benefits of vaccines. That is why we wrote this book.

Here is the first group of secrets you need to know:

1. Vaccines are not safe for every child and the government and pharmaceutical companies have known this for years.
2. Some children will get injured or die from vaccines and the government and pharmaceutical companies know this, too.
3. Pharmaceutical companies have developed an incredible way to make money from vaccines and not be held accountable.

4. When a child is injured or killed by a vaccine, the pharmaceutical company does not pay for the damage it caused—we do!

Fear mongering

Fear sells. Fear works. When people are scared, they do things they might not do when they are thinking clearly.

For example, if a drug company wanted to sell you a vaccine they could say: "This vaccine may help your child avoid a rash, but there are side effects to the vaccine so get all the facts first."

Or, they could say: "There's a public health emergency. If your child doesn't get this vaccine they could end up in a hospital for a long time. And if they don't get vaccinated, they can't go to school."

Which approach do you think would get more people to vaccinate? Which approach would sell more vaccines?

Parents do not want to be bullied into making health decisions for our children. We want the truth and the facts, and we want to make our own decisions for our own families. But, the truth is being hidden. In addition, our right to use our own judgment for our children is also being taken away.

Where the information comes from

This book addresses the misinformation that is being spread by those with power-driven or profit-driven agendas. We reveal the reliable, medically documented information that is being hidden.

All the sources can be found in every chapter so you can check them yourself. We have done our best to cite the most credible sources of information possible, such as the American Academy of Pediatrics, the World Health Organization, and the peer reviewed medical literature where physicians and scientists publish their findings.

So, if you want the facts about the measles vaccine, please read on.

There will be people who are unhappy that we revealed these secrets. To them we say, "Why are you afraid of the truth?"

And one final note to any politician reading this: Before you make policy or laws, please get your facts straight. Do not listen to drug company lobbyists. You are making health decisions for hundreds of thousands of children at once. That is a sacred responsibility. Work for the people who elected you. There are resources in this book that you should be referencing before you make decisions on vaccines. Please do your homework.

PART 1

There Is No Need
to Panic about Measles

We are bombarded with messages that measles is a public health emergency and there are "outbreaks" everywhere. We are told that if we do not vaccinate our children, not only do we put our own children in grave danger, but we jeopardize the lives of everyone else in our communities. But, measles is not as dangerous as we are being told and not as prevalent nor rampant. In this chapter you will learn the facts.

Secrets:
1. Measles Is Usually Just a Rash
2. There's Over a 90 Percent Chance of NO Complications from Measles
3. Measles Is Not as Lethal as We Are Told
4. Fear Mongering Is Used Because It Works

Measles Is Usually Just a Rash

> *Quick Version:* According to pediatricians, measles is basically a rash. It can sometimes be itchy and may cause a stuffy nose and pink eye.

What is measles?

The American Academy of Pediatrics (AAP) is an American professional association of pediatricians. It has 64,000 members and is considered the preeminent organization for pediatricians.

According to the AAP:

> Measles is an acute viral disease characterized by fever, cough, coryza, and conjunctivitis, followed by a maculopapular rash beginning on the face and spreading cephalocaudally and centrifugally.

Sounds pretty horrible, so let's translate this into English:
- Coryza = inflammation of the mucous membrane in the nose. Most people call this a "stuffy nose."
- Conjunctivitis = often called "pink eye," this is a common eye disease, especially in children. It usually lasts a few days and goes away by itself.
- Maculopapular rash = a type of rash that is covered with small bumps. It can sometimes be itchy.

- Cephalocaudally and Centrifugally = moving downward and away from center.

So, what is measles, really?

According to pediatricians, measles is basically a rash. It can sometimes be itchy and spread across the body. It may cause a stuffy nose and pink eye.

Source:

Kimberlin, David W., and Mary A. Jackson. *Red Book 2018: Report of the Committee on Infectious Diseases*, 31st ed. American Academy of Pediatrics, 2018.

There's Over a 90 Percent Chance of NO Complications from Measles

Quick Version: Children who get measles have a more than 90 percent chance of zero complications.

Although measles is basically a rash, we are told that the complications can be serious and sometimes deadly. Is this true?

Let's see what the Centers for Disease Control and Prevention (CDC) said *before* the advent of the vaccine.

In 1962, CDC says measles has low fatality.

The CDC is the leading national public health institute of the United States. The CDC is a United States federal agency under the Department of Health and Human Services.

According to the former head of the CDC, Dr. Alexander D. Langmuir, measles is a "self-limiting infection of short duration, moderate severity, and low fatality . . . " He said this in 1962, *before* the measles vaccine was in use.

The term self-limiting means a disease that tends to go away on its own, without treatment.

According to the National Institute of Health:

Dr. Langmuir was the first chief epidemiologist of the newly established Communicable Disease Center (now the Centers for Disease Control and

Prevention) a position he held for over 20 years . . . [He] was recognized internationally as an assertive public health authority.

In the early 1960s, parents and pediatricians viewed measles as "an unpleasant, although more or less inevitable, part of childhood."

According to an article published in the *American Journal of Public Health*:

> By the early 1960s, almost all children contracted measles before they reached
> adolescence . . . By 1960, thanks to the use of antibiotics and improvements
> in living conditions, measles mortality was declining steadily in industrial-
> ized countries. Parents largely came to see measles as an unpleasant, although
> more or less inevitable, part of childhood . . . Many primary care physicians
> shared this view.

So, measles was not a "big deal" in the years leading up to the measles vaccine and many "physicians shared this view."

The co-inventor of the measles vaccine said most kids have measles "*without* terrible complications."

Here's a story shared by Dr. Samuel Katz about the dangers of measles.
Who is Samuel Katz? He is the co-inventor of the measles vaccine. In inter-
views given in 2002, Katz describes how they tested his vaccine in the 1960s
in a small town in Nigeria called Imesi. They waited until the vaccine was
approved in the United States "because we were very concerned we could be
thought of as using these poor black kids as guinea pigs."

In Nigeria, Katz worked with David Morley, an English physician.
While Katz was at Imesi, Dr. Morley's son got measles. This is what happened in Katz's own words:

> The other thing that was interesting was that while I was there David Morley's
> son got measles. He had his own family there, and we hadn't immunized any
> of the personnel of the hospital. So he had measles like any other child in this
> country having measles. He was sick, got a rash, he coughed, had conjuncti-
> vitis, and he got better . . . Here was this healthy, well-nourished English kid
> who has measles the way hopefully most kids were to have measles, without
> terrible complications or problems.

The co-inventor of the measles vaccine tells us that a well-nourished English kid got measles. He got a rash, coughed, and got pink eye. Then he got better "without terrible complications and problems" like most kids.

What are the complications from measles?

What are we told about the complications from measles today? According to the American Academy of Pediatrics:

> Complications of measles, including otitis media, bronchopneumonia, laryngotracheo-bronchitis (croup), and diarrhea, occur commonly in young children and immunocompromised hosts.

Let's translate that into English:
- Otitis media = ear infection (7 percent of the time)
- Bronchopneumonia = inflammation of the lungs, arising in the bronchi or bronchioles (6 percent of the time)
- Laryngotracheobronchitis (croup) = "barking" cough and a hoarse voice (percentage not provided)
- Diarrhea (8 percent of the time)

Therefore, if your child has measles, they have a 93 percent chance of NOT having an ear infection, a 94 percent chance of NOT getting bronchopneumonia, and a 92 percent chance of NOT getting diarrhea.

Perhaps this is the reason that the first chief epidemiologist of the CDC wrote in 1962 that measles is a "self-limiting infection of short duration, moderate severity, and low fatality."

Perhaps this is also why pediatricians in the early 1960s shared the view with parents that measles was "an unpleasant, although more or less inevitable, part of childhood."

Perhaps this is also why the co-inventor of the measles vaccine, Sam Katz, said that his friend's son got measles like most kids, without complications.

Bronchopneumonia was rare and afflicted children who were malnourished or weakened from another disease.

Of all the complications listed above, bronchopneumonia is considered the most dangerous. This is what was written about bronchopneumonia from measles in 1937, decades *before* the measles vaccine. This comes from the *British Medical Journal*:

> Bronchopneumonia remains the most serious and dreaded complication . . .
> In measles, bronchitis is present in the early stages in all but the mildest and
> modified cases. It usually clears up in a few days; but occasionally, and
> especially in poorly nourished and debilitated children, it progresses to
> bronchopneumonia.

Let's think about what this means:

In 1937, in the years before the vaccine and antibiotics and modern healthcare, and clean running water, measles could sometimes lead to bronchitis and bronchitis could "occasionally" lead to bronchopneumonia in children, especially those who are malnourished or weakened from another disease. Does that describe your child? Is he or she "poorly nourished" or "debilitated"? Are you living in 1937?

Should this scare us today when we have antibiotics and clean running water and children who are not malnourished?

Measles is a rash for the vast majority of children.

Again, for the vast majority of children, measles is basically a rash. In fact, in 1962, before the measles vaccine, the following statement appeared in the *British Medical Journal*: "Measles is often regarded as a normal part of childhood development . . . "

Sources:

Cunningham, A. A. "Vaccine Treatment of Measles." *BMJ* 1, no. 3988 (June 1937), 1202–1203. doi:10.1136/bmj.1.3988.1202.

Hendriks, Jan, and Stuart Blume. "Measles Vaccination Before the Measles-Mumps-Rubella Vaccine." *American Journal of Public Health* 103, no. 8 (2013), 1393–1401. doi:10.2105/ajph.2012.301075.

"Inactivated Measles Virus Vaccine." *BMJ* 1, no. 5294 (June 1962), 1746–1747. doi:10.1136/bmj.1.5294.1746.

Katz, MD, Samuel L. "Oral History Project." Interview by Jeffrey P. Baker, MD, PhD. *American Academy of Pediatrics*, March 2002.

Katz, MD, Samuel L. "Oral History Project." Interview by Jeffrey P. Baker, MD, PhD. *American Academy of Pediatrics*, June 2002.

Kimberlin, David W., and Mary A. Jackson. *Red Book 2018: Report of the Committee on Infectious Diseases*, 31st ed. American Academy of Pediatrics, 2018.

Langmuir, Alexander D. "Medical Importance of Measles." *Archives of Pediatrics & Adolescent Medicine* 103, no. 3 (1962), 224. doi:10.1001/archpedi.1962.02080020236005.

"Measles." Centers for Disease Control and Prevention. Last modified September 25, 2019. https://www.cdc.gov/vaccines/pubs/pinkbook/meas.html.

Schultz, Myron G., and William Schaffner. "Alexander Duncan Langmuir." *Emerging Infectious Diseases* 21, no. 9 (September 2015), 1635–1637. doi:10.3201/eid2109.141445.

SECRET #3

Measles Is Not as Lethal as We Are Told

> **Quick Version:** In the last five years, there have been 2,014 cases of measles and zero deaths in children. There are 250,000 deaths a year from medical errors. The media screams to us about measles, but is silent about medical errors. Why?

The last measles death in the United States was in 2015, but this was an adult with "underlying health problems," not a child.

According to the CDC, "The last measles death in the United States occurred in 2015."

The victim was a twenty-eight-year-old woman "with underlying health problems."

What were these underlying health problems?

According to the *Seattle Times*, "The young woman's health condition was redacted from the public records, but it required her to take drugs that suppressed her immune system."

Whatever she was taking made her vulnerable to falling ill from many different diseases. We don't know if she had cancer or an autoimmune disease (both diseases use chemo to suppress the immune system) or if there were other health issues because that information was redacted.

Why was it redacted? The woman's privacy could have been respected by never revealing her name. Perhaps it was because the information would show that whatever health challenges this unfortunate woman faced, they weren't typical.

The CDC also states, "For every 1,000 children who get measles, one or two will die from it." Their own data, however, does not support this statement.

From 2015 until 2019, there were 2,014 cases of measles and one death. But that death was of a twenty-eight-year-old woman, not a child.

If we use the CDC's own data from the last five years, we have zero children's deaths out of 2,014 cases. That's not one or two deaths out of every thousand cases. That's zero.

Year	Cases	Pediatric Deaths
2015	188	0
2016	85	0
2017	120	0
2018	372	0
2019	1249	0
Total	2014	0

You would think that if measles is as "serious" as we are told that the data on measles deaths would be precise with pediatric deaths separated from adult deaths. It's not.

Many of the recent measles cases were in vaccinated people.

Let's look at the most recent data from 2019.

According to the CDC's Morbidity and Mortality Weekly Report Early Release, there were 1,249 reports of measles cases in 2019. The CDC states:

- 142 of the 1,249 measles cases were in people *who were vaccinated*.
- 235 of the 1,249 of the measles cases may have been vaccinated because their vaccination status was unknown.
- Sixty of the 1,249, or 5 percent, had pneumonia, and one (0.1 percent) had encephalitis.
- No deaths were reported to CDC.

So, no deaths out of 1,249 cases and 0.1 percent had encephalitis (one case out of 1,249).

Remember, we are talking about the risk of death from measles in the United States. In other countries where there are problems with nutrition, clean water, general hygiene, and good medical care, measles does represent a greater risk, but so does almost every other disease.

Let's put measles in perspective.

Back in the United States, we need some perspective about measles. Based on CDC data we now know that out of the last 2,014 cases of measles in the last five years, there has not been a single death of a child.

Let's compare that to:

- 250,000 deaths per year due to medical errors in the United States
- 2,167 deaths from constipation every year
- Three hundred bystanders and passengers are killed as a result of high-speed police chases annually
- Twenty thousand people die each year from vigorous exercise
- Texting while driving kills six thousand annually

As you can see, there are many things that are much more dangerous than measles.

250,000 deaths every year from medical errors

There are 250,000 deaths every year from medical errors, which is horrendous. This figure comes from a 2016 Johns Hopkins study. It's equivalent to two jumbo jets crashing every single day, 365 days a year. And this has been going on for years. A million deaths in just the last four years. Where is the non-stop news coverage about this? How many of the victims are children? Where are the breaking news reports? Have you seen any? Seems strange that there is silence about a million deaths in the last four years while the media whips itself into a frenzy about measles with zero deaths in the same amount of time.

Without the measles vaccine, there still would be few measles deaths.

Some people might say that if it weren't for the measles vaccine, the measles death rate would be much higher, perhaps as high as constipation.

This is what Dr. Samuel Katz, the co-inventor of the measles vaccine, wrote in 1960 (before the measles vaccine was in use): "[T]he morbidity rate

of uncomplicated [measles] alone is staggering in terms of school absence..."
In other words, when kids get measles they stay home from school. He
didn't say they were dying.

Let's take a look at the actual numbers because that data exists.

The measles vaccine was first introduced in 1963, but it wasn't widely
used until 1967 when a campaign was launched to try to eliminate measles
from the United States. In 1968, the US Department of Health, Education,
and Welfare published *Vital Statistics Rates in the United States 1940–1960.*
This is the most reliable source of statistical information about diseases in
the United States.

According to its report, in 1966, there were 261 deaths from measles in
the entire country. That may sound like a lot, so let's put it in perspective.

In 1966, there were 196.6 million people in the United States. That
means the measles death rate was .00000132.

Measles deaths were disappearing before the vaccine.

In the next graph, you can see the death rate from measles from 1900–1960.
This is government data. It is their graph. Remember that the vaccine was
introduced in 1963 and in widespread use starting in 1967. Notice how the
line was going down many years before the vaccine.

Figure 19.—Death Rates for Measles: Death-registration States,
1900–32, and United States, 1933–60

(Rates per 100,000 population)

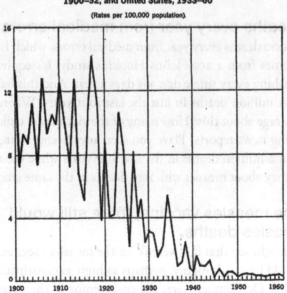

The media would have you believe that if it weren't for the vaccine, measles and measles deaths would be rampant in the United States. This is simply not true.

Why would they keep this a secret? Why not tell the public that death from measles was very small *before* the measles vaccine was rolled out? Isn't this important to know?

Sources:

Aleccia, JoNel. "Fatal measles case linked to exposure at tribal clinic, records show." The Seattle Times. Last modified February 29, 2016. https://www.seattletimes.com/seattle-news/health/fatal-measles-case-linked-to-exposure-at-tribal-clinic-records-show/.

"Complications of Measles." Centers for Disease Control and Prevention. Accessed February 8, 2021. https://www.cdc.gov/measles/about/complications.html.

Katz, Samuel L. "Studies on an Attenuated Measles-Virus Vaccine." *American Journal of Diseases of Children* 100, no. 6 (July 1960), 942. doi:10.1001/archpedi.1960.04020040944023.

Kimberlin, David W., and Mary A. Jackson. *Red Book 2018: Report of the Committee on Infectious Diseases*, 31st ed. American Academy of Pediatrics, 2018.

Kane, Sean. "8 Surprisingly Common Causes of Death That Sound Like Freak Accidents." *Business Insider*. Last modified July 18, 2016. https://www.businessinsider.com/weird-causes-of-death-2016-7.

Makary, M. A., and M. Daniel. "Study Suggests Medical Errors Now Third Leading Cause of Death in the U.S.—05/03/2016." Johns Hopkins Medicine, Based in Baltimore, Maryland. Accessed February 8, 2021. https://www.hopkinsmedicine.org/news/media/releases/study_suggests_medical_errors_now_third_leading_cause_of_death_in_the_us.

"Measles Data and Statistics." Centers for Disease Control and Prevention. Accessed February 8, 2021. https://www.cdc.gov/measles/downloads/MeaslesDataAndStatsSlideSet.pdf.

Patel, Manisha, Adria D. Lee, Nakia S. Clemmons, Susan B. Redd, Sarah Poser, Debra Blog, Jane R. Zucker, et al. "National Update on Measles Cases and Outbreaks—United States, January 1–October 1, 2019." *MMWR. Morbidity and Mortality Weekly Report* 68, no. 40 (October 2019), 893–896. doi:10.15585/mmwr.mm6840e2.

Patel, Manisha. "National Update on Measles Cases and Outbreaks — United States, . . . " Centers for Disease Control and Prevention. Last modified October 9, 2019. https://www.cdc.gov/mmwr/volumes/68/wr/mm6840e2.htm#T1_down.

"Study Suggests Medical Errors Now Third Leading Cause of Death in the U.S.—05/03/2016." Johns Hopkins Medicine. Last modified May 3, 2006. https://www.hopkinsmedicine.org/news/media/releases/study_suggests_medical_errors_now_third_leading_cause_of_death_in_the_us.

"Underlying Cause of Death, 1999–2018 Request." CDC WONDER. Accessed February 8, 2021. https://wonder.cdc.gov/ucd-icd10.html.

SECRET #4

Fear Mongering Is Used Because It Works

> *Quick Version:* The media, the government, and doctors with financial interests in vaccines have become very skilled in scaring us to get what they want. They ignore the medical facts and use a highly effective "propaganda playbook."

Fear mongering is used to convince parents to vaccinate their children. If you watch TV, you have seen it in action.

Do you remember the Disneyland measles "outbreak" a few years ago? This was used as an excuse to change the laws in California and force more children to be vaccinated.

What about the so-called US measles epidemic of early 2019? Hundreds of hours were taken up on network broadcasts. Thousands of articles in the mainstream media were devoted to the same subject. Fear was spread on a daily basis. How many people died? Zero. But, that doesn't matter. What mattered was that vaccine laws were changed and that exemptions were taken away in New York and other states.

When exemptions go away, it means more children will be vaccinated. More children mean more customers for the hundred plus new vaccines in the pharmaceutical industry's pipeline.

This is not new. Fear has been used to scare parents into vaccinating their children for decades.

Hepatitis B, is this leading or misleading?

Here's an example from twenty years ago. You will see that the "propaganda playbook" has been used for other vaccines, not just for measles and it's been around a long time.

The doctor who co-invented the measles vaccine is Samuel Katz. In 1999, he went on national TV to spread a little fear about hepatitis B and why every child needed a hepatitis B vaccine.

On October 14, 1999, Katz appeared on a show called *Nightline* which airs on ABC. On that show he stated:

> Hepatitis B infection in infants and children leaves an incredibly high risk of chronic liver disease, of cirrhosis, of cancer of the liver. There are four or five thousand individuals a year in our country who die of the complications of hepatitis B.

Pretty scary. But, Katz made no mention that children born of healthy parents are at an infinitesimal risk of contracting hepatitis B.

He did not say that Dr. George Peter, chairman of the American Academy of Pediatrics, stated at the National Pediatric Infectious Disease Seminar on June 12, 1992 in Washington, DC that one reason to recommend hepatitis B vaccine to infants is because "children are accessible."

He did not mention that, in the United States, the rates of acute hepatitis B infection have remained about one per hundred thousand since 2009.

He didn't state that hepatitis B is a risk for intravenous drug abusers and prostitutes who have unprotected sex, not children.

He didn't explain that most hepatitis B infections clear up within one to two months without treatment.

He did not mention the study *Reactions of Pediatricians to a New Center for Disease Control Recommendation for Universal Immunization of Infants with Hepatitis B Vaccine* that was published in journal Pediatrics in 1993 that surveyed 778 pediatricians of whom "only 32% of those questioned believed that hepatitis B vaccination was warranted in their practices."

He also did not mention the adverse effects listed by the vaccine's manufacturer, Merck, as including: abdominal pains, influenza, vertigo, myalgia, earache, dysuria, hypotension, Guillain-Barre Syndrome, multiple sclerosis, myelitis, peripheral neuropathy such as Bell's Palsy, radiculopathy and visual disturbances.

But, Katz did imply that without the hepatitis B vaccine, your child might get liver cancer.

That's how fear mongering works. It's not just what is said, it's what is not said.

In an interview in 2002, Katz did admit, "[H]epatitis in young children by and large was not a severe infection. It might lead to chronic illness, but as far as the acute infection went, adults have severe hepatitis, while children often have occult infection or fairly benign infection."

Study shows 99.999 percent of babies either did not have a Hepatitis B infection or it went away.

Here's an example of real data about hepatitis B from the medical literature. This study was published in the *Journal of Pediatrics*.

- 4,453 pregnant women were examined
- Twenty-five of the 4,453 were carriers of hepatitis B **(that's 0.62 percent, less than 1 percent).**
- So 99.38 percent did not have a hepatitis B infection.
- Sixteen infants from the twenty-eight hepatitis B positive moms were studied.
- Sixteen did not have a current hepatitis B infection, but eight later did **(that's 0.18 percent, much less than 1 percent).**
- By the time the study was written, the hepatitis infection had already gone away in seven of the eight infants.
- **So 99.999 percent of babies either did not have a hepatitis B infection or it went away. That's the actual math.**
- There is no mention that any child had chronic liver disease, cirrhosis, or liver cancer.

A balanced presentation of the risk of hepatitis B in children in the United States would have included data like this, but it was left out.

Katz also said, "There are four or five thousand individuals a year in our country who die of the complications of hepatitis B." This is also not true.

According to the US Department of Health and Human Services National Vital Statistics Reports, for the year 1999 (when Katz was making these claims), there were a total of 4,853 deaths related to ALL hepatitis types including: A, B, and C combined, not just hepatitis B.

Of those 4,853 deaths, a total of eight were in children (ages zero to fourteen years of age) and only two were in infants. Remember this was for all types of hepatitis.

A balanced presentation would have shared numbers like these. But, of course, all this was left out too.

With the Hepatitis B vaccine, there are actually more cases of Hepatitis B.

Here's another interesting fact: In 1997, the CDC began recommending the hepatitis B vaccine for children. The rationale was that it would protect kids when they were adults because some of them may be at higher risk by becoming IV drug addicts, prostitutes, etc. Here are some data for all hepatitis cases from the US Department of Health and Human Services.

Year	Number of Hepatitis Deaths
1996 (no hepatitis B vaccine)	3,780
2017 (most recent data available after twenty years of the hepatitis B vaccine)	5,611

Today, there are more hepatitis B deaths with the vaccine than without!

Yes, the population has increased from about 269 million in 1996 to 325 million in 2017, but the hepatitis death rate has increased even more. In other words, even with hepatitis B vaccination, there are more deaths from hepatitis B today than before the vaccine.

But, on the bright side, the drug companies Merck and GlaxoSmith Kline have sold at least ten billion dollars' worth of hepatitis vaccines during the same period.

The Measles "Propaganda Playbook" uses three steps to change the laws.

With measles, the propaganda playbook has become much more sophisticated. A three step program is used against parents:

- Step one: spread fear and hyperbole. Scare us.
- Step two: change the law so people are forced to vaccinate. Take away our rights.

- Step three: Once the law is changed, assure the public that everything is fine and that the scary outbreak and epidemic didn't really mean anything. Everything goes back to normal.

Playbook in action—the New York example
Let's look at how this propaganda playbook was recently used in New York.

Step one: Fear and hyperbole. Scare us.
Here were some of the headlines from New York in the first half of 2019. Do you remember any of these?

- *New York City declares a public health emergency amid Brooklyn measles outbreak*
- *New York Declares Health Emergency As Measles Spreads In Parts Of Brooklyn*
- *Measles nears record in US as the disease spreads in New York*
- *N.Y. Suburb Declares Measles Emergency, Bars Unvaccinated Minors From Public Places*

Now that people are scared we have . . .

Step two: Change the law so people are forced to vaccinate. Take away our rights.
On June 13, 2019, New York Governor Andrew Cuomo signed legislation that removes non-medical exemptions from school vaccination requirements. The law went into effect immediately. This means that the religious and philosophical exemptions that normally allow people to say "no" to one or more vaccines were taken away.

Now that the law was changed and vaccines are essentially mandatory, we move on to . . .

Step three: Quietly assure the public that everything is fine and that the scary outbreak and epidemic didn't really mean anything. Everything is back to normal.
On October 4, 2019, Health and Human Services Secretary Alex Azar said, "We are very pleased that the measles outbreak has ended in New York and that measles is still considered eliminated in the United States."

What? Wait a second. Outbreaks and public health emergencies are being screamed at us with headlines like "*Measles nears record in US as the disease spreads* . . . "and yet measles is still eliminated from the United States? Huh?

The authorities and drug companies gain more power over your child's body.

Once the law was changed, the government tells us that everything is fine.

But, things are not fine.

First, this propaganda was used to take away your decision-making power. The law was changed. Now, you cannot decide whether or not your child will undergo vaccination, a medical intervention. It's not up to you anymore. It is now up to politicians and their donors, the drug companies.

Second, measles is not eliminated. It has never been eliminated. There have been 3,279 cases of measles over the last ten years in the United States.

But, the truth doesn't seem to matter. What matters is that the law was changed and children are now being forced to be vaccinated.

Who would want this? Perhaps the companies that sell billions of dollars of vaccines every year, spend millions on the media, and contribute many more millions to the politicians that change these laws. What do you think?

One thing is for sure. Fear works.

Human interest stories are an important part of the playbook.

If you really want to capture people's attention and scare them, human interest stories are most effective.

The media has perfected this type of fear-based reporting. This is just one example. It was published by CNN on May 6, 2019. It was called "This mom wants you to know what measles did to her baby." The article was designed to scare parents.

The story describes Alba, who is an eleven-month-old baby. Her mom is named Jilly Moss. Alba contracted measles and ended up in the hospital. But, the story wasn't that simple.

From March 25 to April 6, 2019, the doctors diagnosed Alba with tonsillitis. She was not treated for measles. The article says, "The doctors, since they'd never seen measles before, misdiagnosed her repeatedly, sending her home, where she became sicker and sicker."

Finally, after almost two weeks of delay and misdiagnosis, Alba was diagnosed with measles and treated with Vitamin A (the standard treatment for measles).

Alba was discharged from the hospital after eight days. She never needed to go to the intensive care unit.

What was interesting about this story was that the two weeks of misdiagnosis were not discussed as one of the reasons Alba got very sick. It is a strange omission.

If your child gets sick and the doctors get it wrong for two weeks you can assume things may get worse. The article says the doctors "misdiagnosed her repeatedly, sending her home, where she became sicker and sicker." But, conveniently, the importance of the doctors' mistakes were ignored.

An appropriate and timely treatment would have saved this child a lot of misery.

Another part of the story that does not appear in the first few paragraphs is that Alba and her family live in Southwest London, England.

CNN had to go thousands of miles away to find the "nightmare" described in the story. This is a story about a British family, not an American one.

If measles is such a threat here in the United States, surely a similar story could be found here?

That's how fear mongering works. Find scary pieces of information wherever you can. Amplify this information. Don't present a reasoned view and leave important information out.

Online sites and internet trolls

The final part of the propaganda playbook is to finance online organizations and internet trolls who will attack anyone who questions compulsory vaccination as an "anti-vaxxer."

We are told, falsely, that diseases are making a comeback due to anti-vaxxers who refuse to vaccinate their kids. Anti-vaxxers are described as "anti-science" home school moms, "rich people," or corrupt doctors.

The truth is that most parents who don't believe in compulsory vaccination are people whose children have already been harmed by vaccines or who have researched the issue in-depth.

The truth is also that there are many doctors who know that the benefits of vaccination are exaggerated and the risks are minimized. Some of these doctors have spoken out because they actually care about protecting children's health.

Nonetheless, the attacks on well-meaning doctors and conscientious parents have the desired effect of chilling debate and discussion. Being labeled as an "anti-vaxxer" is another fear tactic used in today's propaganda playbook.

Sources:

Bahler, Kristen. "Rich People Are Leading the Anti-Vaccine Movement—and Experts Have a Theory Why." *Money.* Last modified April 15, 2019. https://money.com/money/5641663/anti-vaccine-rich-people/.

"Dangerous Vaccinations?" *Nightline.* ABC, October 14, 1999.

Dupuy, J. M. "Hepatitis B in Children. II. Study of Children Born to Chronic HBsAg Carrier Mothers." *Pediatrics* 92, no. 2 (February 1978), 200–204.

Freed, G. L. "Reactions of Pediatricians to a New Center for Disease Control Recommendation for Universal Immunization of Infants with Hepatitis B Vaccine." *Pediatrics* 91, no. 4 (April 1993), 699–702.

Grove, Robert D. "Vital Statistics Rates in the United States 1940–1960." *Centers for Disease Control and Prevention.* Accessed February 14, 2021. https://www.cdc.gov/nchs/data/vsus/vsrates1940_60.pdf.

"Hepatitis B Symptoms & Treatment in Children | Children's Pittsburgh." UPMC Children's Hospital of Pittsburgh. Accessed February 9, 2021. https://www.chp.edu/our-services/transplant/liver/education/liver-disease-states/hepatitis-b.

Hoyert, Donna L. *National Vital Statistics Reports.* Hyattsville, MD: Center for Disease Control, 2001.

Interview by Jeffrey P. Baker, MD, PhD. *American Academy of Pediatrics,* March 2002.

Katz, MD, Samuel L. "Oral History Project." Interview by Jeffrey P. Baker, MD, PhD. *American Academy of Pediatrics,* June 2002.

LaVito, Angelica. "Measles Nears Record in US As the Disease Spreads in New York." CNBC. Last modified April 22, 2019. https://www.cnbc.com/2019/04/22/measles-nears-record-in-us-as-the-disease-spreads-in-new-york.html.

National Vital Statistics Reports, Number of Deaths from Selected Causes, by Age. United States: U.S. Department of Health and Human Services, 2017.

Nelson, Noele P., Philippa J. Easterbrook, and Brian J. McMahon. "Epidemiology of Hepatitis B Virus Infection and Impact of Vaccination on Disease." *Clinics in Liver Disease* 20, no. 4 (November 2016), 607–628. doi:10.1016/j.cld.2016.06.006.

News Staff. "In Wake of Measles Outbreaks, CDC Updates 2019 Case Totals—U.S. Keeps Measles Elimination Status." American Academy of Family Physicians. Last modified October 9, 2019. https://www.aafp.org/news/health-of-the-public/20191009measlesupdt.html.

Perry, Tod. "Anti-science Mother Argues Her Kids Are Better off Homeschooled Than Getting Immunized." GOOD. Last modified August 23, 2019. https://www.good.is/anti-science-mom-mad-school-sign.

Peter, George. Presentation, National Pediatric Disease Seminar, Washington, DC, June 12, 1992.

Romo, Vanessa. "New York Declares Health Emergency As Measles Spreads In Parts Of Brooklyn." NPR.org. Last modified April 9, 2019. https://www.npr.org/2019/04/09/711432792/new-york-declares-health-emergency-as-measles-spreads-in-parts-of-brooklyn.

Schwartz, Matthew S. "N.Y. Suburb Declares Measles Emergency, Bars Unvaccinated Minors From Public Places." NPR.org. Last modified March 27, 2019. https://www.npr.org/2019/03/27/707095754/ny-suburb-declares-measles-emergency-bars-unvaccinated-minors-from-public-places.

Scutti, Susan. "New York City Declares a Public Health Emergency Amid Brooklyn Measles Outbreak." CNN Digital. Last modified April 9, 2019. https://www.cnn.com/2019/04/09/health/measles-new-york-emergency-bn/index.html.

Elizabeth Cohen. "This mom wants you to know what measles did to her baby" CNN Digital. Last modified May 6, 2019. https://www.cnn.com/2019/05/06/health/measles-baby-misdiagnosis-eprise/index.html.

"Top 25 Pharma & BioPharma—GlaxoSmithKline Top Selling Drugs." Pharmaceutical and Biopharmaceutical Contract Servicing & Outsourcing—Contract Pharma. Accessed February 9, 2021. https://www.contractpharma.com/heaps/view/6072/3/318091.

PART 2

The Measles Vaccines Are Worth Billions of Dollars to Vaccine Makers and the Government

Why is it important to know about the money side of vaccines when making the decision whether or not to vaccinate your child? For vaccine manufacturers and their shareholders, their primary interest is to sell their product. Media campaigns that promote their product mean larger profits. People who don't use their product mean less profit. And there is A LOT of profit at stake. But, that's not all. The governmental organizations that approve and mandate vaccines also profit from the very vaccines they are supposed to regulate.

Secrets:
5. The Measles Vaccine (and Chickenpox) Is Worth $3.2 Billion *in Only Eighteen Months* to Its Manufacturer (Merck)
6. The Vaccine-Media Connection Where Everyone Profits
7. The FDA and CDC and the People Who Work There Make Money from Vaccines (Sad, But True!)

The Measles Vaccine (and Chickenpox) Is Worth $3.2 Billion *in Only Eighteen Months* to Its Manufacturer (Merck)

> **Quick Version:** The measles vaccine generates billions of dollars for its manufacturer, Merck. This company is associated with price fixing, criminal charges, whistleblower lawsuits, overbilling, data falsification, and wrongful death lawsuits.

The vaccine business has a great money-making model. Here's why:

You have a company. You want to sell your product. The government declares that everyone in the country must buy your product. Jackpot! And if that is not enough, you make sure that you are not financially liable if anyone gets hurt using your product. No one can sue you. So, it's all profit, no risk. A winning business model!

This is the business model for vaccines that companies like Merck enjoy. Merck makes the measles vaccine. But, Merck is not the only vaccine maker. Others include:

- GlaxoSmithKline with annual revenue of more than thirty-five billion dollars.

- Sanofi is another of the top vaccine manufacturers in the world. Its annual revenue is more than forty billion dollars.
- Pfizer is regarded as one of the top vaccine companies with annual revenue around fifty billion dollars.

Merck, measles, money

Merck is a large pharmaceutical company with annual revenue of about forty-three billion dollars. It is headquartered in Kenilworth, New Jersey and operates in more than 140 countries. In other countries it is called "MSD."

Kenneth C. Frazier is Merck's chairman of the board and chief executive officer. He is also a director of Exxon Mobil and PhRMA, a trade and lobbying group representing the pharmaceutical industry.

Before becoming CEO, Frazier was general counsel of Merck and oversaw the company's legal defense against claims that their drug Vioxx caused heart attacks and strokes.

David Graham, who worked in the FDA's Office of Drug Safety, estimated that *sixty thousand people died from Vioxx.* Frazier defended Merck against many of these victims. Later he became its CEO. After causing so many deaths, it's incredible that this company is still in existence.

Frazier makes millions from Merck. The following is from Wikipedia:

> Frazier received a total compensation of $21,387,205 in 2014; $17,023,820 in 2015; and $21,781,200 in 2016. On February 26, 2017, it was reported Frazier owned 600,304 shares of Merck stock worth approximately $37,000,000. Based on stock transactions at Merck alone and his tenure at the company, his net worth is in the hundreds of millions.

That's just the money he makes. His company makes even more money—just from vaccines!

In the eighteen months between April 2018 and September 2019, Merck has sold *$3.2 billion worth of measles and chickenpox vaccines.*

The measles and chickenpox vaccines earn billions for Merck, and business was good after the "epidemic" came and went.

- MMRII is the measles, mumps, and rubella vaccine.
- ProQuad is the measles, mumps, rubella, and varicella (chickenpox vaccine).
- Varivax is just the chickenpox vaccine.

All three are made by Merck. Here is what Merck published in its quarterly earnings reports for MMRII, ProQuad, and Varivax:

- 2018 Q2—$426,000,000
- 2018 Q3—$525,000,000
- 2018 Q4—$455,000,000
- 2019 Q1—$496,000,000
- 2019 Q2—$675,000,000
- 2019 Q3—$623,000,000

Total: $3,200,000,000

On October 29, 2019, after the so-called measles "epidemic" erupted and then quietly vanished after vaccine exemptions were taken away, Merck proudly announced that its' "Human Health Vaccines Sales Grew 17%." In fact you can see the increase in sales at the end of 2018 when the measles "epidemic" was being screamed from the rooftops.

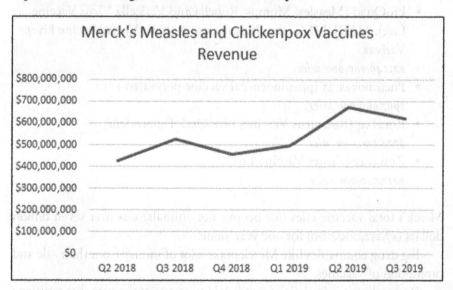

The current cost of one dose of MMR is $75.04. Assuming the average cost of just forty dollars a dose since 1978, that means Merck has grossed **eight billion dollars from the MMR vaccine alone in the United States and eighteen billion dollars worldwide.**

Vaccines are "substantial products" for Merck.

According to Adbrands, "the worlds' leading advertisers and agencies," vaccines are an essential part of Merck's business.

In 2010, they wrote:

> Merck & Co. regained a position among the world's leading pharmaceutical companies. . . . The company has made a dynamic recovery . . . turning what appeared to be a clutch of minor vaccines into what are now *substantial products*. . . . Other blockbusters include the ProQuad MMR vaccine.

Remember, ProQuad is the measles, mumps, rubella, and varicella (chickenpox) vaccine.

Here is why they say the vaccines are "substantial products" for Merck. These numbers are for one year (2018) and come from Merck's Annual Report:

- Gardasil Vaccine
  $3,151,000,000 sales
- ProQuad (Measles, Mumps, Rubella and Varicella Virus Vaccine Live), MMRII (Measles, Mumps and Rubella Virus Vaccine Live), Varivax
  $2,249,000,000 sales
- Pneumovax 23 (pneumococcal vaccine polyvalent)
  $907,000,000 sales
- RotaTeq (Rotavirus Vaccine, Live Oral, Pentavalent)
  $728,000,000 sales
- Zostavax (Zoster Vaccine Live)
  $217,000,000 sales

Merck's total vaccine sales (for people, not animals) was over seven billion dollars ($7,252,000,000) for one year alone.

Big drug companies like Merck make a lot of money from their sale and promotion of vaccines.

Remember, it is in their commercial interest to make sure that everyone uses their product. But, to grow their business even bigger they need to sell more vaccines.

Do you think that laws forcing people to buy their product would be good for business?

The World Health Organization says vaccines will be bigger than ever.

According to the World Health Organization:

- As of 2019, there were more than 120 new vaccines in the development pipeline.
- The vaccine market has increased in value from five billion dollars in 2000 to almost twenty-four billion in 2013.
- The global vaccine market is projected to rise to a hundred billion dollars by 2025.
- Vaccines are becoming an engine for the pharmaceutical industry.
- Newer and more expensive vaccines are coming onto the market faster than ever before. In fact, Merck is trying to bring an Ebola vaccine to market in the next couple of years. (They call it V920.)
- Five large multinational corporations make up 80 percent of the global vaccine market.

Merck has made billions from the measles vaccine and many more billions from other vaccines. Its company leaders have been paid many millions of dollars. They have more vaccines in the pipeline. They want every American to be required to take their vaccines because it means immense profits.

Do you think the measles vaccine and other vaccines would be pushed so hard if they didn't make any money?

But the problem isn't only the pursuit of money. The other problem is that companies like Merck, who are trusted because we inject their products into our children, should not be trusted.

Can Merck be trusted?

Here are some examples of Merck's troubles with price fixing, criminal charges, whistleblower lawsuits, overbilling, data falsification, and wrongful death lawsuits.

These are the actions where Merck got caught. Is it responsible for other illegal, unethical, or unsafe activities where it hasn't yet been caught? Only Merck knows.

As you read these examples, you can decide for yourself if the government is doing a good job of policing the company that injects substances directly into your child.

Perhaps the question should be asked—should a company with this many problems be in the vaccine business at all?

Merck has had troubles with: price fixing, criminal charges, whistleblower lawsuits, overbilling, data falsification, and wrongful death lawsuits.

1996

Merck was one of fifteen drug companies that paid more than $408 million to settle a class action lawsuit charging that they conspired to fix prices charged to independent pharmacies.

1997–2001

Merck maintained numerous sales programs that disguised excess payments to physicians as compensation for "training," "consultation," or "market research." One might call these "bribes."

2001

Merck's drug Vioxx increased the risk of heart attack or stroke. Internal company e-mails revealed that Merck's management appeared to have known about the risks of Vioxx for years. In 2007, the company agreed to pay $4.85 billion to pay people injured or killed by their drug. They also paid $950 million to resolve criminal charges and agreed to plead guilty to violating federal drug law and pay an additional $321.6 million criminal fine. As described above, David Graham, who worked in the FDA's Office of Drug Safety, estimated that sixty thousand people died from Vioxx.

2007

Merck agreed to pay $1.6 million in penalties to settle charges relating to Clean Water Act violations at its plant in Montgomery County, Pennsylvania.

2008

Merck agreed to pay the federal government more than $650 million to settle charges that the company routinely overbilled Medicaid and other government programs and made illegal payments to healthcare professionals to induce them to prescribe its products.

2008

Merck agreed to pay $671 million to federal and state prosecutors for allegedly overcharging government programs for four drugs—Zocor, Mevacor, Vioxx, and Pepcid, and for bribing doctors to prescribe certain drugs. This is one of the largest health care fraud settlements to date.

2010

Stephen Krahling and Joan Wlochowski, former Merck virologists, claim that they "witnessed firsthand the improper testing and data falsification in which Merck engaged to artificially inflate [the mumps component of] the (MMR) vaccine's efficacy findings." According to the complaint:

> This case is about Merck's efforts for more than a decade to defraud the United
> States through Merck's ongoing scheme to sell the government a mumps vac-
> cine that is mislabeled, misbranded, adulterated and falsely certified as having
> an efficacy rate that is significantly higher than it actually is . . . .

2010

Twenty-one-year-old Christina Richelle Tarsell died after receiving Merck's Gardasil vaccine. Christina's family fought a long eight-year legal battle with the government-run vaccine court, and the US Government eventually conceded that the Gardasil vaccine killed twenty-one-year-old Christina. However, the compensation received by Christina's family came out of the National Vaccine Injury Compensation Program, funded by taxes collected from the sale of vaccines. Merck suffered no consequences because US law protects them.

2011

The Massachusetts attorney general announced that Merck would pay twenty-four million dollars as its part of a forty-seven-million-dollar settlement reached with thirteen drug makers to resolve allegations that they overcharged the state's Medicaid program.

2012

The Louisiana attorney general announced that Merck and four other companies would pay a total of $25.2 million to resolve allegations that they overcharged the state's Medicaid program.

Current Case

In an attempt to prevent shingles, some patients may have unintentionally contracted the virus from their shingles vaccination. Lawsuits are now being filed for injuries allegedly caused by Merck's Zostavax vaccine.

Is the US Government doing a good job of regulating this company? Should this company be in the vaccine business? Remember, this is the company that you trust with your children's lives.

Sources:

"CDC Vaccine Price List." Centers for Disease Control and Prevention. Last modified August 28, 2020. https://www.cdc.gov/vaccines/programs/vfc/awardees/vaccine-man agement/price-list/index.html.

Herper, Matthew. "David Graham On The Vioxx Verdict." *Forbes*, August 19, 2005. https://www.forbes.com/2005/08/19/merck-vioxx-graham_cx_mh_0819graham.html?sh=81c34c25698e

"Investor Relations." Merck.com. Last modified February 9, 2021. https://investors.merck.com/financials/quarterly-reports/default.aspx.

"Kenneth Frazier." Wikipedia, the Free Encyclopedia. Last modified December 1, 2010. https://en.wikipedia.org/wiki/Kenneth_Frazier.

"Merck Announces Third Quarter Financial Results." Merck. Last modified 2019. https://s21.q4cdn.com/488056881/files/doc_financials/2019/q3/Merck-3Q19-Earnings-News-Release.pdf.

"Merck & Co Advertising & Marketing Assignments." Adbrands.net. Accessed February 9, 2021. https://www.adbrands.net/us/merck_us.htm.

"United States v. Merck & Co. Pennsylvania Eastern District Court, Case No. 2:10-cv-04374-CDJ." *Keller Grover.* 2012. https://www.dropbox.com/s/81ruzsot1loy9lr/Merck%20Amended%20ComplaintECFStamped.pdf?dl=0.

The Vaccine-Media Connection
Where Everyone Profits

> **Quick Version:** The media is not objective about vaccines because they receive billions of dollars in advertising revenue from the drug companies that make vaccines. It wouldn't be smart to bite the hand that feeds them. Instead of taking a hard look at the vaccine industry and vaccine safety, they act as cheerleaders and scaremongers.

If you think you are seeing more drug commercials on TV, you are not alone.

Direct to consumer advertising is on the rise.

When you see drugs being advertised on TV this is called "Direct to Consumer Advertising," or DTC.

The US is one of only two countries that permit drug makers to advertise their products directly to consumers (the other is New Zealand). And their favorite way to do that has been through TV.

DTC is a fairly recent phenomenon. It has been legal in the United States since 1985, but only became popular in 1997 when the FDA relaxed rules that required companies to offer a detailed list of side effects in their long format television commercials.

Since then, DTC advertising has more than quadrupled. Money spent on DTC (mostly TV commercials) went from $2.1 billion in 1997 to $9.6 billion in 2016.

$43.3 billion spent on TV advertising

Who benefits from all this drug company spending? The big TV broadcasters and cable companies.

According to the media agency Magna, $43.3 billion was spent on all TV advertising in 2016 (not just from pharmaceutical companies).

This suggests that around 22 percent of all the money that the mass media receives comes from the pharmaceutical industry ($9.6 billion out of $43.3 billion). And this percentage continues to increase.

Does that mean that a cable or broadcast station will work with its advertisers to promote stories that the advertisers want? Of course it does. This is called "editorial support" and it's been around a long time.

For example: "A 1992 US study of 150 newspaper editors found that 90 percent said that advertisers tried to interfere with newspaper content and 70 percent said that advertisers tried to stop news stories altogether." Forty percent admitted that advertisers had in fact influenced a story.

Interlocking boards connect media and the drug companies.

It's not just advertisers using their influence to change media coverage. It's more than that. Many of the same people who sit on media executive boards and management boards also are connected to drug companies. For example, a recent study by the organization Fairness & Accuracy in Reporting (FAIR) found deep connections between drug companies like Merck and the big media companies on TV. They wrote:

> [A] crossover between these media corporations and several large pharmaceutical companies, such as Eli Lilly, Merck and Novartis . . . Out of the nine media corporations studied, six had directors who also represented the interests of at least one pharmaceutical company. In fact, save for CBS, every media corporation had board connections to either an insurance or pharmaceutical company.

How often do you see a TV investigative report that presents the connection between the media and the pharmaceutical industry? Never. Why would they reveal their own secret?

How often does the TV news say something negative about a drug or vaccine? Hardly ever. Does this mean that all drugs and vaccines are perfect? Definitely not. FDA-approved drugs and vaccines have been responsible for tens of thousands of deaths.

This all means that, unfortunately, you can't count on the media to present a fair and objective view of vaccines. It's not smart to bite the hand that feeds you.

Sources:

http://www.herinst.org/BusinessManagedDemocracy/government/media/Adinfluence .html citing: Andre Carothers, "The Green Machine," New Internationalist (August 1993), p. 16.

Michael F. Jacobson and Laurie Ann Mazur, Marketing Madness (Boulder, Colorado: Westview Press, 1995), p. 207.

"The Green Machine." *New Internationalist (August 1993), p. 16;.* August 1993.

Jacobson, Michael F., and Laurie A. Mazur. *Marketing Madness,* 207. Boulder, CO: Westview Press, 1995.

Bell, Jacob. "Pharma Advertising in 2018: TV, Midterms and Specialty Drugs." BioPharma Dive. Last modified September 26, 2018. https://www.biopharmadive.com/news/pharma -ad-dtc-marketing-2018-spend-TV-congress/533319/.

Dunn, Andrew. "Pharma DTC Spending Outpaces Rest of Medical Marketing, JAMA Study Finds." BioPharma Dive. Last modified January 7, 2019. https://www.biopharmadive .com/news/pharma-dtc-spending-outpaces-rest-of-medical-marketing-jama-study -finds/545441/.

"Influence of Advertisers on News." Her Institute. Accessed February 9, 2021. https://www .herinst.org/BusinessManagedDemocracy/government/media/Adinfluence.html.

Lafayette, Jon. "TV Revenue Expected to Dip in Record-Setting Year for Ads." Broadcasting Cable. Last modified September 20, 2018. https://www.broadcastingcable.com/news /tv-revenue-expected-to-dip-in-record-setting-year-for-ads.

Mole, Beth. "Big Pharma Shells out $20B Each Year to Schmooze Docs, $6B on Drug Ads." Ars Technica. Last modified January 11, 2019. https://arstechnica.com/science/2019/01 /healthcare-industry-spends-30b-on-marketing-most-of-it-goes-to-doctors/.

Murphy, Kate. "Single-Payer & Interlocking Directorates: The Corporate Ties Between Insurers and Media Companies." FAIR. Accessed February 9, 2021. https://fair.org /home/single-payer-and-interlocking-directorates/. July 1, 2009.

The FDA and CDC and the People Who Work There Make Money from Vaccines (Sad, But True!)

> *Quick Version:* The Department of Health and Human Services (HHS) is over the Food and Drug Administration (FDA) and the Centers for Disease Control (CDC). All three organizations and the people within them make money from vaccines.

The agencies and people who approve and make vaccines mandatory for the United States *have received money* from vaccine manufacturers like Merck. In turn, they help Merck sell more vaccines.

Think about this and how this corrupts what you are told about vaccines.

Have you seen that reported on CNN or Fox or any other news station screaming about "states of emergency" from measles? Where are the reports that show how the biggest vaccine proponents are the same people and organizations who are financed by vaccines?

Let's start at the beginning:

HHS is over the NIH, FDA, and CDC.

The Department of Health and Human Services (HHS) is a Cabinet Level Department in the Federal Government. HHS has a $1.2 trillion dollar budget. It is over the National Institute of Health (NIH), the CDC, and the FDA.

The FDA approves vaccines and the CDC puts them on the schedule for children in the entire country, making them mandatory.

Then the CDC buys billions of dollars' worth of vaccines. In fact, the CDC buys about half of all childhood vaccines in the United States through the Vaccines for Children (VFC) Program. They then sell them to contracted public health agencies. So the CDC makes vaccines mandatory and also buys them. We will return to this in a moment.

HHS—It stinks at the top.
The former head of HHS was appointed by President Trump. His name is Alex Michael Azar II. He's a good example of the revolving door between industry and government.

- On August 3, 2001, Azar was confirmed to be the general counsel of the United States Department of Health and Human Services.
- He resigned in January 2007.
- Six months later, in June 2007, he was hired by Eli Lilly and Company to be the company's top lobbyist and spokesman as its senior vice president of Corporate Affairs.
- On January 1, 2012, Azar became president of Lilly USA, LLC, the largest division of Eli Lilly and Company, and was responsible for the company's entire operations in the United States.
- Prices for drugs rose substantially under Azar's leadership.
- Azar also served on the board of directors of the Biotechnology Innovation Organization, a pharmaceutical lobby.
- This lobbying company includes every major vaccine maker including Merck, GlaxoSmithKline, Sanofi, and Pfizer. Last year, this drug and vaccine lobbying organization spent $9,870,000 to lobby the government.
- Azar has previously been on the Board of Directors of the Healthcare Leadership Council, where he was treasurer. This organization has, among other things, wanted a complete corporate takeover of the Medicare program. Lobbyists work for this group as well and members of this organization include vaccine makers like Merck and Pfizer.
- In January 2017, Azar resigned from Eli Lilly. He also resigned from the Board of Directors of the Biotechnology Innovation Organization.

- Eleven months later, on November 13, 2017, President Trump announced via Twitter that he would nominate Azar to be the next United States Secretary of Health and Human Services.

In summary, the person who sat at the top of HHS has worked for drug companies and has led lobbying organizations whose members are the biggest vaccine makers. Now, the organizations under his control (like the CDC and FDA) decide what vaccines are mandatory.

HHS and NIH get paid by vaccine makers.

Next, let's see how these vaccine makers actually pay HHS.

Another organization under HHS is the National Institute of Health (NIH). NIH uses taxpayer money to fund research. When that research leads to something that can be sold, NIH patents it and then licenses the patent to companies like Merck. HHS then gets a cut of the profits.

This means that the government can create a vaccine with our tax money, license it to a drug company, approve it, and make it mandatory. Once it does that, it ensures more profits for itself.

This is called self-dealing. It is usually illegal, but in the world of vaccines, it's not.

For example, the HPV vaccines Gardasil and Cervarix came from research patented by the NIH's National Cancer Institute (NCI). The NCI then licensed the technology to Merck, MedImmune, and GlaxoSmithKline. By 2009, HPV licensing had become NIH's top generator of royalty revenues.

As you will see, vaccines are a gravy train and all the government entities are lapping up the profits.

CDC gets paid by vaccine makers.

The CDC decides which vaccines are put on the pediatric vaccine schedule and therefore what is mandatory throughout much of the United States.

The CDC is part of our government, but it is also in the vaccine business. Under a 1980 law, the CDC currently has twenty-eight licensing agreements with companies and one university for vaccines or vaccine-related products. It also has eight ongoing projects to collaborate on new vaccines.

A search conducted of vaccine-related patents through "Google Patents" delivered 168 results for CDC as assignee or applicant.

Vaccines for Children (VFC) buys billions of dollars-worth of vaccines.

In addition, the CDC is like an ATM for the vaccine manufacturer. Here's how:

Under the Vaccines for Children (VFC) Act, funding for new vaccines or new recommendations occurs through a vote of the CDC's Advisory Committee on Immunization Practices (ACIP). When ACIP votes to add a vaccine to the schedule it can also have it covered by VFC.

This means that this one committee (ACIP) has the power and authority to add benefits to an entitlement program. This is very unusual.

After ACIP passes a VFC resolution for a vaccine, the federal government establishes a contract with the relevant manufacturer. Funding is approved by HHS and the Office of Management and Budget *without* the need for additional Congressional appropriations.

So ACIP is essentially the gateway by which vaccine makers make multi-billion-dollar sales!

Not surprisingly, members of ACIP have received money from vaccine manufacturers. Relationships have included: sharing a vaccine patent; owning stock in a vaccine company; receiving payments for research; getting money to monitor manufacturer vaccine tests; and funding academic departments.

Let's connect some dots.

A vaccine manufacturer like Merck can use its considerable financial power to influence the people who sit on the ACIP committee. This committee can then add a new vaccine to the schedule making it mandatory for much of the whole country.

But that's not all. Through VFC, this committee can write a purchase order for billions of dollars' worth of vaccines from the same manufacturer (i.e. Merck).

Merck-CDC connections

Are there really connections between Merck and ACIP? Yes, here are some examples.

In 2001, the House Government Reform Committee looked into the CDC's Vaccine Advisory Committee. This is the committee that recommends whether or not to make a vaccine mandatory (add it to the vaccine schedule).

Their August 2001 report found that "four out of eight CDC advisory committee members who voted to approve guidelines for the rotavirus vaccine in June 1998 had financial ties to pharmaceutical companies that were developing different versions of the vaccine."

Even a quick look at the current members of the CDC's Vaccine Advisory Committee is revealing.

One member is Dr. Kevin Ault. He has received money from Merck in the past. In fact, Dr. Ault served as a clinical trials investigator for Merck's Gardasil vaccine during his tenure at University of Iowa School of Medicine.

Gardasil is a vaccine that is supposed to prevent cervical cancer. In its clinical testing it never prevented one case of cervical cancer. Not one! Also, according to VAERS there have been 51,641 injuries and 241 deaths associated with the vaccine. Even the US government has admitted that Gardasil kills in the Christina Tarsell case.

Another member of the CDC's Vaccine Advisory Committee is Dr. Peter Szilagyi. He has received over ten million dollars from CDC to research vaccines.

Another member is Dr. David S. Stephens. He has served on review panels for the NIH, FDA, and CDC. He also has vaccine-related patents. Yet he is sitting on committees that will decide which vaccines may be mandatory for every child in the country.

FDA gets paid by vaccine makers.

Before a vaccine is put on the vaccination schedule by the CDC it first has to be approved by the FDA. However, the FDA actually receives and depends on money from companies like Merck in order to function.

Yes, that's true. In 1992, the drug industry started paying the salaries of drug reviewers.

"The virginity was lost in '92," said Dr. Jerry Avorn, a professor at Harvard Medical School. "Once you have that paying relationship, it creates a dynamic that's not a healthy one."

According to an investigative report from ProPublica:

- In 2017, pharmaceutical companies paid 75 percent ($905 million) of the agency's scientific review budgets for branded and generic drugs, compared to 27 percent in 1993.
- Industry also sways the FDA through a less direct financial route. Many of the physicians, caregivers, and other witnesses before FDA advisory panels that evaluate drugs receive consulting fees, expense payments, or other remuneration from pharmaceutical companies.

- "You know who never shows up at the [advisory committee]? The people who died in the [clinical] trial," lamented one former FDA staffer, who asked not to be named because he still works in the field. "Nobody is talking for them."

Another investigation from *Science* was entitled "Investigation Examines Big Pharma Payments to FDA Advisers." It found:

- Forty of 107 physician advisers on the committees examined "received more than $10,000 in post hoc earnings or research support from the makers of drugs that the panels voted to approve, or from competing firms," the publication said. Of those 40, 26 snagged more than a hundred thousand dollars and seven of those gained one million dollars or more.
- *Science* said that seventeen top-earning advisers benefitted from more than twenty-six million dollars in research assistance or personal payments from industry companies.
- Yale University physician Robert Steinbrook, editor at large for *JAMA Internal Medicine*, told *Science* that such payments are "troubling" and raises ethical concerns.

Conflicts of interest at the FDA
The FDA was not doing its due diligence when it approved the rotavirus vaccine ("RotaShield").

In June 2000, the Committee on Government Reform looked at the approval of this vaccine. It was made by Merck and approved for use by the FDA in August 1998. It was recommended for universal use by the CDC in March 1999. Children were soon injured or killed by it.

Congress found:

- Members of the two committees, including the chair of the FDA and CDC advisory committees who made the vaccine approval decisions, owned stock in Merck. Merck was the drug company that made the vaccine.
- Individuals on both advisory committees owned patents for vaccines that were affected by the decisions of the committees.
- Three out of the five of the members of the FDA's advisory committee who voted for the rotavirus vaccine had conflicts of interest that were waived.

The tangled web of money and influence that corrupts vaccine science and decision-making

Let's recap. We now know:

- HHS is over the CDC, FDA, NIH, and NCI. It was run by a person with ties to vaccine makers.
- NIH uses taxpayer money to fund research. When that research leads to something that can be sold, NIH patents it and then licenses the patent to companies like Merck. HHS then gets a cut of the profits.
- The CDC is in the vaccine business because of its financial licensing interests in various vaccine products and its vaccine patents.
- The CDC's Advisory Committee on Immunization Practices (ACIP) strongly influences which vaccines are mandatory for the country (which vaccines are put on the pediatric vaccine schedule). Members of this committee have financial and other connections to vaccine makers.
- The CDC actually buys billions of dollars of vaccines under the Vaccines for Children (VFC) Act.
- The FDA gets paid by vaccine makers.
- Members of the FDA's vaccine committee have financial and other connections to vaccine makers.

Sadly, we are not done yet.

CDC foundation gets sponsored by Merck.

Most people have heard of the CDC, but almost no one has heard of the CDC Foundation.

What is the CDC Foundation?

On paper, the CDC Foundation is separate from CDC. It is a private, nonprofit 501(c)(3) organization. It was created to help the CDC work with the private sector. But the reality appears to be something different. For example, Dr. Judith Monroe is the president and CEO of the CDC Foundation. Before that, she was deputy director of the CDC.

In addition, both the CDC and the CDC Foundation are getting money from the same place that the CDC does. Big drug companies like Merck.

For example, in 2017 the CDC Foundation's budget was $86,096,884. One of the sponsors of the organization was Merck. How much of that eighty-six million dollars came from this one drug and vaccine company?

They do not report it. The CDC and the CDC Foundation are both in Atlanta and are about four miles apart.

Dr. Judith Monroe's relationship with Merck goes back to at least 2009 when she was president of an organization called US Association of State and Territorial Health Officials (ASTHO), and Merck was one of its sponsors.

Let's continue following Merck's influence in the vaccines injected into your children.

Here are more glaring examples of how vaccine decision making has been corrupted by money.

Julie Gerberding: From director of the CDC to Merck

Julie Gerberding actually ran the CDC from 2002 to 2009. Her title was Director of US Centers for Disease Control and Prevention. During her leadership there, the CDC approved a number of vaccines, including Merck's Gardasil. In late 2009, Gerberding left the CDC and became president of Merck's vaccines division. At Merck, she is reportedly receiving a multi-million-dollar compensation package. Gardasil sales in 2018 were $3.15 billion dollars.

Dr. Brenda Fitzgerald buys and sells Merck stock while director of CDC.

Dr. Brenda Fitzgerald was the director of the CDC between July 2017 and January 2018. She had to resign after it was reported that she bought tens of thousands of dollars in new stock holdings in at least a dozen companies—including Merck.

Dr. Paul Offit's Merck connection

Dr. Paul Offit is the doctor who helped approve RotaShield (the rotavirus vaccine that injured and killed children). He owned a patent on another rotavirus vaccine that he licensed to Merck and has reportedly been paid millions of dollars.

He also directs the Vaccine Education Center at the Children's Hospital of Philadelphia (CHOP). And he holds the Maurice R. Hilleman Chair in Vaccinology.

He was a member of ACIP. ACIP is CDC's Advisory Committee on Immunization Practices. This is the group of medical and public health experts that develop vaccine recommendations for every child in the country by deciding what goes on the pediatric vaccine schedule.

Some of the money he and his university have received comes directly from Merck. For example:

- Maurice R. Hilleman was a former senior vice president of Merck. As the Maurice R. Hilleman Chair in Vaccinology, a $1.5 million endowment goes to Offit's University. How much he receives of that is unclear.
- At ACIP, Offit voted three times in favor of decisions related to the use of the rotavirus vaccine (RotaShield). At this time, he shared ownership of a patent for a rotavirus vaccine being developed under a grant from Merck. RotaShield was later pulled off the market after it injured and killed children.
- Offit's own vaccine was approved by the FDA in 2006 under the trademark "RotaTeq." Offit profited. He stated that he made "several million dollars, a lot of money." According to one investigative report, "Offit had earned $10 million in royalties through 2009 and stood to gain anywhere from $3–25 million in additional payments, depending on the commercial performance of the RotaTeq franchise."

Here's another estimate of the money Dr. Offit made from this ONE vaccine:

1. The Children's Hospital of Philadelphia payout from the Royalty Pharma sale— $6.2 million.
2. Royalty payouts from CHOP prior to the Royalty Pharma sale— $0.5 million.
3. The Wistar payout from the Paul Capital Royalty Fund—$2.3 million.
4. Royalty payouts from the royalty stream that Wistar did not sell to Paul Capital—$1.0 million.
5. Offit's payout from the sale of his Wistar royalty stream—$7.5 million est. ($3–20 million range).
6. As if that's not enough money for Offit, Merck bought and delivered copies of Offit's book, *What Every Parent Should Know About Vaccines*, to American doctors. The book has a list price of $14.95.
7. Offit said he does not know how many copies of his book Merck purchased. "I don't have any control over that," he said.

Do you think that people like Offit and Gerberding should make millions of dollars from vaccines and also have a role in deciding what vaccines are mandatory for your child? Isn't this wrong?

In summary, there are conflicts of interest between the makers of vaccines like Merck and the government approval organizations like the FDA and the CDC. These conflicts extend to the organizations themselves and to individuals who profit personally.

Vaccine approval should be done dispassionately and objectively. It's a medical intervention for healthy children so there should be an abundance of caution that goes into its review. But, it's not happening. Too much money is at stake and companies like Merck are very good at paying to get their way.

Sources:

Ault, Kevin A. "Human papillomavirus infections: diagnosis, treatment, and hope for a vaccine." *Obstetrics and Gynecology Clinics of North America* 30, no. 4 (December 2003), 809–817.

Benjamin, Mark. "UPI Investigates: The Vaccine Conflict." UPI. Last modified July 21, 2003. https://www.upi.com/Odd_News/2003/07/21/UPI-Investigates-The-vaccine-conflict /44221058841736/?ur3=1.

Blaxill, Mark. "Offit Cashes In: Closing the Books on the Vaccine Profits of a Merck-Made Millionaire." Age of Autism. Last modified January 31, 2011. https://www.ageofautism .com/2011/01/offit-cashes-in-closing-the-books-on-the-vaccine-profits-of-a-merck -made-millionaire-1.html.

Caceres, Marco. "What Would Jesus Do About Measles?" The Vaccine Reaction. Last modified June 2, 2015. https://thevaccinereaction.org/2015/06/what-would-jesus-do -about-measles/.

Chen, Caroline. "FDA Repays Industry by Rushing Risky Drugs to Market." ProPublica. Last modified June 26, 2018. https://www.propublica.org/article/fda-repays-industry-by -rushing-risky-drugs-to-market.

"Close Ties and Financial Entanglements: The CDC-Guaranteed Vaccine Market." Children's Health Defense. Last modified July 15, 2019. https://childrenshealthdefense .org/news/close-ties-and-financial-entanglements-the-cdc-guaranteed-vaccine-market/.

"Corporations, Foundations & Organizations—Fiscal Year 2017 Report to Contributors." CDC Foundation. Accessed February 10, 2021. https://www.cdcfoundation.org/FY2017 /organizations.

"Corporations, Foundations & Organizations—Fiscal Year 2018 Report to Contributors." CDC Foundation. Accessed February 10, 2021. https://www.cdcfoundation.org/FY2018 /organizations.

"Executive Branch Personnel Public Financial Disclosure Report: Periodic Transaction Report (OGE Form 278-T)." POLITICO. Last modified December 21, 2017. https://www.politico.com/f/?id=00000161-4804-d9fe-a9fd-5af5834d0000.

"FDA Increasingly Approves Drugs Without Conclusive Proof They Work." PBS NewsHour. Last modified June 26, 2018. https://www.pbs.org/newshour/health/fda-increasingly-approves-drugs-without-conclusive-proof-they-work.

Hayes, Carol. "ACNM Liaison Meeting Report—CDC Advisory Committee on Immunization." American College of Nurse-Midwives. Last modified October 2019. https://www.midwife.org/acnm/files/ccLibraryFiles/Filename/000000005743/ACNM-Liaison-Mtg-Rpt-CDC-ACIP-Oct-2015.pdf.

Hinman, A. R., W. A. Orenstein, and L. Rodewald. "Financing Immunizations in the United States." *Clinical Infectious Diseases* 38, no. 10 (May 2004), 1440–1446. doi:10.1086/420748.

"Inventor David S. Stephens." Google Patents. Accessed February 10, 2021. https://patents.google.com/?inventor=David+S+Stephens.

Keown, Alex. "Investigation Examines Big Pharma Payments to FDA Advisers." BioSpace. Last modified July 6, 2018. https://www.biospace.com/article/investigation-examines-big-pharma-payments-to-fda-advisers/.

PART 3
The Measles Vaccine Can Be Dangerous to Some Children

The measles vaccine, like every other vaccine, can be dangerous or even deadly to some children. The evidence for this is overwhelming. It shouldn't be surprising. Vaccines are medical interventions like prescription drugs. Have you ever known anyone to have a reaction to a prescription they received from their doctor? It's the same with vaccines. One-size-fits-all means some children will be injured. But, it's worse with vaccines because they do not undergo the same safety testing as drugs.

Secrets:
8. The Measles Vaccine Was Never Safe Since the Beginning
9. Safety Testing of the Measles Vaccine Was Totally Inadequate
10. The National Academy of Medicine Says the Measles Vaccine Can Cause Febrile Seizures, Anaphylaxis, and Arthralgia in Some Children
11. The National Academy of Medicine Says There Is NOT Enough Information to Determine if the Measles Vaccine Is Completely Safe
12. The Measles Vaccine and All Vaccines Are Considered Legally "Unavoidably Unsafe"
13. The Measles Vaccine Maker Admits There Are Over Seventy Potential Side Effects

14. According to the CDC and FDA, There Are 89,032 Cases of Injury Associated with the Measles Vaccine, or Is It a Hundred Times as Much?
15. The US Government Has Children Injured by the Measles Vaccine in Their Database
16. The VAERS Database Says There Are 275 Deaths Associated with the Measles Vaccine
17. Many Doctors Admit in Their Publications that the Measles Vaccine Can Cause Dangerous Side Effects
18. Doctors Admit that Vaccines Can Cause Autoimmune Diseases
19. Safety Is in the Eye of the Beholder

SECRET #8

The Measles Vaccine Was Never Safe Since the Beginning

Quick Version: Studies done on the measles vaccine almost sixty years ago revealed that it caused neurological side effects and other problems in some children.

There were neurological side effects, seizures, fever, and rash from the measles vaccine since the beginning.

The measles vaccine was tested in the late 1950s and early 1960s. The results of these tests were published and they revealed that children were getting sick from this vaccine since the very beginning. Even in its early tests, the measles vaccine resulted in bronchitis, seizures, cyanosis, and aberrations of encephalographic patterns caused by a neurologic insult of some type.

For example, a study was published in 1962 about 345 children who received the new measles vaccine. Here's what the study said:

> Fevers of 100°F or more were recorded for 79% of the children, with 3.1 days the mean duration of fever . . . A rash was reported in 52% of the histories . . .
>
> Three instances of convulsive seizure following vaccination were reported in these studies, two of them by the same pediatrician . . .
>
> One three-year old girl (C.J.) experienced a seizure on the seventh post-vaccination day and was immediately hospitalized. Her temperature was

found to be 101, and rose to 103 on the following day and to 104 on the third day, when a rash appeared. No respiratory signs were observed.

Electroencephalographic examinations on admission and those made two weeks later revealed patterns which were reported by the neuropathologist to be unlike those seen in encephalitis, but were suggestive of an epileptogenic area in the centroencephalon.

(Translation: This child had a neurological side effect to the vaccine.)

A daughter of the pediatrician who had vaccinated C.J. two months earlier was the second child to develop a seizure (S.S.). She was twenty months old, and had experienced a similar seizure prior to vaccination. The postvaccination seizure was associated with a temperature of 104 on the evening of the seventh day after inoculation. A brief tonic seizure accompanied by cyanosis was followed by a generalized clonic seizure lasting about two minutes. On the following day the child's temperature rose to 104.8 and a rash appeared. Her subsequent history was uneventful.

(Translation: Tonic seizures involve sudden stiffening and contraction of the muscles. Clonic seizures involve rhythmic twitching or jerking of one or several muscles. Both types of seizures happened after getting the vaccine.)

A 14-month-old child (T.C.) was the third to experience a seizure. Like the other two seizures described, this one occurred on the seventh postvaccination day. The child was promptly examined by a pediatrician who reported a temperature of 104 . . . Low fever persisted for four to five days and a rash was noted for three days . . . Fevers of 103° to 104.90 were reported for only 6 per cent of the home dwelling children, but for 21 per cent of those in an orphanage . . . Perhaps it is an overstatement to say that in our preoccupation with febrile responses we are measuring differences in children, not differences in vaccine.

Remember, this was written by doctors who were testing the measles vaccine about sixty years ago. These statements reveal some disturbing facts:

1. When the measles vaccine was first being tested, there were already side effects being observed including fevers, rash, and seizures.
2. There were also neurologic side effects to the measles vaccine. But, despite this, the doctors focused more on the fevers (what they call "febrile responses").
3. The doctors understood that the reactions these children were having to the vaccine were due to the fact that all children are not the same. The researchers said that they were "measuring differences in children."

More neurological side effects

These observations should have given these doctors pause to reconsider what they were doing. After all, if there were neurologic side effects, that by itself was a red flag. Understanding that not all children will react the same way to the same vaccine was another important insight that was ignored. And getting seizures after being vaccinated is not a good thing.

In another study, one child out of twenty-eight had a "nonspecific aberration of her encephalographic pattern." Here's what the co-inventor of the measles vaccine wrote:

> Gibbs et al have administered this [measles] vaccine to 28 susceptible children of whom at least 17 to date have responded clinically . . . Electroencephalograms obtained during the vaccine response was completely normal in 27 and unusual in only 1 child who on the fifth postvaccinal day had an intercurrent infection and presented a nonspecific aberration of her encephalographic pattern.
>
> *(Translation: In this tiny study, one child out of twenty-eight had a neurological reaction five days after getting the vaccine.)*

This was yet another clue that the measles vaccine could cause neurologic side effects in some children.

In that same study the researchers wrote, "Moderate fever and circumscribed rash, however, may be regarded as desirable reactions to measles vaccine since they present easily recognized criteria of successful vaccination."

So, if the vaccine causes fever and rash it's a good thing because that indicates that the vaccine worked?

Bronchitis caused by the vaccine

In another study, the reactions consisted of bronchitis, but that didn't seem to matter either.

Here's what the researchers found:

Eighty-eight children who did not have measles antibodies were given the vaccine. They ranged in age from one to fifteen. Fifty-three percent were under five and 82 percent were under eight.

- Sixty-five of the eighty-eight children had fever with a maximum temperature of 106°F.
- Thirty-three got a rash.
- Three got an ear infection ("mild otitis media").

- Two got bronchitis.

Most children were observed for only fourteen days. Ten children were observed eight to twelve months later. Apparently, only their measles antibody levels were checked, not their general health.

Why the rush to stop a rash?

All of these early studies demonstrate that the scientists ignored the potential for dangerous side effects and were in a rush to bring the measles vaccine to market.

In fact, an article in the *American Journal of Public Health* said that vaccine development in the United States, in the mid twentieth century, was "marked by a current of urgency."

This "current of urgency" allowed scientists to ignore danger signals and plow straight ahead, focusing only on one small aspect of vaccination—did the children have a reaction that would indicate an antibody response to the vaccine? If we have "moderate fever and circumscribed rash" then we are doing well because it means there was reaction and we have a "successful vaccination."

Such blindness isn't excused because of the scientists' current of urgency. What was the rush? For most children, measles was a rash. It was not considered a clear and present danger by most pediatricians and parents at that time.

Big questions go unanswered since the beginning.

This scientific urgency didn't answer the pressing questions about the measles vaccine:

- Why were there neurological side effects from the vaccine?
- What exactly did the vaccine do in the bodies and brains of different children?
- How did the vaccine permeate different tissues of the body?
- Why did the vaccine cause bronchitis and seizures?
- What exactly did the vaccine do in the nerve cells or intestinal cells of some children?

None of those questions were asked because they were too difficult to answer, especially in the late 1950s. Instead, the scientists ignored what was inconvenient, like neurological reactions, and focused on the easy questions like, "Did we see a rash or not?"

Unfortunately, none of these important signals gave these scientists pause and it was full steam ahead with a vaccine that we still have today. This poorly tested, unsafe, sixty-year-old measles vaccine is still being injected into children as you read this.

Sources:

Hendriks, Jan, and Stuart Blume. "Measles Vaccination Before the Measles-Mumps-Rubella Vaccine." *American Journal of Public Health* 103, no. 8 (August 2013), 1393–1401. doi:10.2105/ajph.2012.301075.

Karelitz, Samuel. "Measles Vaccine." *JAMA* 177, no. 8 (August 1961), 537–541.

Katz, Sameul L. "Studies on an Attenuated Measles-Virus Vaccine." *American Journal of Diseases of Children* 100, no. 6 (July 1960), 180–184. doi:10.1001/archpedi.1960.04020040944023.

Markham, Floyd S., Herald R. Cox, and James M. Ruegsegger. "A Summary of Field Experience with Live Virus Measles Vaccine." *American Journal of Public Health and the Nations Health* 52, no. Suppl_2 (February 1962), 57–64.

SECRET #9

Safety Testing of the Measles
Vaccine Was Totally Inadequate

Quick Version: Vaccines are promoted as safe, but the gold standard of safety testing is not used and long-term side effects are not tracked.

The vaccine industry tells us that all childhood vaccines have been tested and they are safe.

For example Dr. Karen Midthun was director of the Office of Vaccines Research and Review in the FDA's Center for Biologics Evaluation and Research. This is the group the recommends the FDA approval of vaccines. She testified in front of Congress that "vaccines are safe and effective."

Besides ignoring the overwhelming evidence to the contrary, the inadequate testing of vaccines almost assures that there will be problems.

In the previous chapter, we saw how the measles vaccine already caused serious side effects in its early tests, but it was approved anyway. This was probably inevitable as the vaccine wasn't even tested as well as most drugs.

Gold standard not used to test vaccines
When pharmaceutical drugs are tested, the gold standard used is called "double-blind placebo controlled."

What does that mean?

It means that when the drug is tested, one group of patients or subjects will get the new drug and another group will get a placebo like saline or a sugar pill, and neither the doctor nor the test subjects know who is getting what. It's done this way to ensure that there is no bias in the results.

With vaccines, this gold standard is not followed. It wasn't followed in the case of the measles vaccine, and it hasn't been followed with any vaccine since.

First, the doctors know who is getting the vaccine and who isn't so it's not double blind.

Second, most vaccine tests today don't use a true placebo. Instead they may use an adjuvant like mercury or aluminum that is in the vaccine. In other words, the placebo group doesn't get saline or a sugar pill. They get the same harsh toxic ingredients that are also in the vaccine.

Why would they do this?

Because the adjuvant, the toxic ingredient like mercury or aluminum, often causes a reaction. Since both groups get the adjuvant (one group only gets the adjuvant and the other group gets the adjuvant in the vaccine) then both groups have similar reactions and side effects. This lets the vaccine researchers get away with scientific dishonesty. They can say the vaccine is safe because both groups had the same side effects so the vaccine is not responsible.

Of course, if they used a true placebo like saline instead of aluminum then the results would be much different and the vaccines would be shown to be much more dangerous.

Small numbers used to test vaccine

In addition, the number of children used in these vaccine tests is very small. For example, there are about forty-eight million children in the United States under the age of eleven. Here are some examples of the small numbers of children tested with the measles vaccine:

303 children
Studies on an Attenuated Measles-Virus Vaccine N Engl J Med. 1960.

296 children
"Inactivated measles virus vaccine. III. A field trial in young school children" JAMA 1962.

273 children
"Development and evaluation of the Moraten measles virus vaccine" JAMA 1968.

72 children
"Measles immunity after revaccination: results in children vaccinated before 10 months of age." Pediatrics. 1982.

The total number of children tested with the measles vaccine was far less than ten thousand. However, even assuming that ten thousand was the number, it only represents 0.02 percent of the current population of children in this country. In other words, if you test a vaccine in ten thousand children, 99.9 percent of children are not being represented.

Does it make sense to mandate a medical intervention like a vaccine for healthy children when it hasn't been tested on literally 99.9 percent of the population?

Will you pick up rare side effects with testing small numbers of children? Remember this is a vaccine designed to prevent a rash in most children. Shouldn't its safety be assured?

And if you do pick up signals of dangerous side effects (such as neurologic side effects or seizures, etc.), shouldn't you investigate? Shouldn't you use long-term studies that follow children for years to make sure that your anti-rash vaccine isn't causing permanent brain damage?

Shouldn't there be a proper cost-benefit analysis?

Today, things are worse. We have many vaccines that have not been proven safe for large numbers of children, yet they are mandatory.

In summary:

1. Vaccines are not tested using the gold standard used in drug testing.
2. True placebos are not used so that side effects can be ignored.
3. Long-term side effects are not measured because there is no long-term follow-up.
4. Small numbers of children are tested so side effects that may only affect a small percentage of the population are missed.
5. As we saw in the previous chapter, when side effects are discovered and described, they are often ignored.

Sources:

Hendriks, Jan, and Stuart Blume. "Measles Vaccination Before the Measles-Mumps-Rubella Vaccine." *American Journal of Public Health* 103, no. 8 (August 2013), 1393–1401. doi:10.2105/ajph.2012.301075.

Karelitz, Samuel. "Measles Vaccine." *JAMA* 177, no. 8 (August 1961), 537–541.

Katz, Sameul L. "Studies on an Attenuated Measles-Virus Vaccine." *American Journal of Diseases of Children* 100, no. 6 (July 1960), 180–184. doi:10.1001/archpedi.1960.04020040944023.

Markham, Floyd S., Herald R. Cox, and James M. Ruegsegger. "A Summary of Field Experience with Live Virus Measles Vaccine." *American Journal of Public Health and the Nations Health* 52, no. Suppl_2 (February 1962), 57–64.

Winkelstein, W., et al., "Inactivated measles virus vaccine. III. A field trial in young school children" JAMA 1962 Feb 10;179:398–403.

M. R. Hilleman, et al., "Development and evaluation of the Moraten measles virus vaccine" JAMA 1968 Oct 14;206(3):587–90.

C. C. Linnemann Jr, et al., "Measles immunity after revaccination: results in children vaccinated before 10 months of age." Pediatrics. 1982 Mar;69(3):332–5.

"Dan Burton Questions Officials at Thimerosal Hearings." *Vaccine Truth*. Accessed February 15, 2021. https://vaccinetruth.org/dan-burton.html.

SECRET #10

The National Academy of Medicine Says the Measles Vaccine Can Cause Febrile Seizures, Anaphylaxis, and Arthralgia in Some Children

Quick Version: The world-renowned National Academy of Medicine says that the measles vaccine can cause injuries in some children.

The Institute of Medicine (now called the National Academy of Medicine) is the preeminent scientific institution in the country.

According to Wikipedia:

> The National Academy of Medicine provides national and international advice on issues relating to health, medicine, health policy, and biomedical science. It aims to provide unbiased, evidence-based, and authoritative information and advice concerning health and science policy to policy-makers, professionals, leaders in every sector of society, and the public at large.

The National Academy of Medicine discovers dangerous side effects with the measles vaccine.

In September 1993, the Institute of Medicine released a report entitled "Adverse Events Associated With Childhood Vaccines: Evidence Bearing on Causality." The report concluded with the following statements:

> The committee found that the evidence favored acceptance of a causal relation between measles vaccine and anaphylaxis.
>
> *(Translation: MMR causes a potentially life threatening allergic reaction.)*
> The committee found that the evidence established causality between measles vaccine and death from measles vaccine-strain viral infection . . . [and between] measles-mumps-rubella vaccine and thrombocytopenia and anaphylaxis.
>
> *(Translation: MMR causes a condition in which you have a low blood platelet count.)*

In 2011, more side effects were reported. In that year, the Institute of Medicine published a report titled "Adverse Effects of Vaccines: Evidence and Causality."

Their vaccine report reviewed over 130 scientific studies on the measles vaccine and concluded:

- "The evidence convincingly supports a causal relationship between MMR vaccine and measles inclusion body encephalitis in individuals with demonstrated immunodeficiencies."
  (Translation: MMR causes a kind of encephalitis – inflammation of the brain in children who are immunodeficient.)
- "The evidence convincingly supports a causal relationship between MMR vaccine and febrile seizures."
  (Translation: MMR vaccine causes seizures with fever.)
- "The evidence convincingly supports a causal relationship between MMR vaccine and anaphylaxis."
  (Translation: MMR causes a potentially life threatening allergic reaction.)
- "The evidence favors acceptance of a causal relationship between MMR vaccine and transient arthralgia (joint pain) in children."

Each time these leaders in science and medicine looked at the measles vaccine they found that it caused severe reactions and injury for some children.

Anyone who claims that the measles or MMR vaccine is completely safe is either ignorant or lying.

Sources:

Stratton, Kathleen R. "Adverse Events Associated With Childhood Vaccines Other Than Pertussis and Rubella." *JAMA* 271, no. 20 (May 1994), 1602–1605.

Stratton, Kathleen R. "Adverse Events Associated with Childhood Vaccines Other Than Pertussis and Rubella. Summary of a Report from the Institute of Medicine." *PubMed*. Last modified August 25, 2011. https://www.ncbi.nlm.nih.gov/pubmed/8182813.

The National Academy of Medicine Says There Is NOT Enough Information to Determine if the Measles Vaccine Is Completely Safe

Quick Version: The National Academy of Medicine also says there isn't enough research to determine if the measles vaccine can cause a long list of other problems. FDA approval does not insure the measles vaccine or any drug or vaccine is safe.

The National Academy of Medicine (formerly the Institute of Medicine) recognized that the measles vaccine can cause severe reactions and injuries. The Academy went on to examine other injuries, problems, side effects and diseases that could be attributed to the vaccine.

The National Academy of Medicine cannot rule out other side effects.

The report said, "The **evidence is inadequate** to accept or reject a causal relationship between MMR vaccine and . . . " (emphasis added)

- fibromyalgia

- chronic fatigue syndrome
- hepatitis
- brachial neuritis
- chronic inflammatory disseminated polyneuropathy
- multiple sclerosis
- neuromyelitis optica
- transverse myelitis
- acute disseminated encephalomyelitis
- meningitis
- encephalopathy
- and other diseases and disorders.

If the vaccine was proven safe, and it was scientifically verified that the vaccine did not cause these other diseases and disorders, the report would have ruled them out. But the report would not say that!

The report suggested that more studies are needed to determine if the vaccine does or does not cause these additional problems. But, the studies are not being done, so the children who get the vaccine are the guinea pigs. Our kids are supplying the data. They are part of a mass experiment. And with compulsory vaccination, you have no choice.

FDA approval does not mean a pharmaceutical or vaccine is safe.

Remember that just because a vaccine is "FDA approved" does not mean it is safe. **For example, many "FDA-approved" drugs have killed thousands of Americans.** Here are some examples:

Accutane (Isotretinoin) caused increased risk of birth defects, miscarriages, and premature births when used by pregnant women. It also caused inflammatory bowel disease and suicidal tendencies. This drug was FDA approved.

Baycol (Cerivastatin) caused serious cardiovascular adverse events (like death, heart attack, and stroke). It also increased risk of serious skin reactions (like toxic epidermal necrolysis, Stevens-Johnson syndrome, and erythema multiforme) and gastrointestinal bleeding. This drug was FDA approved.

Darvon & Darvocet (Propoxyphene) caused serious toxicity to the heart. Between 1981 and 1999 there were over 2,110 deaths reported. This drug was FDA approved.

DES (Diethylstibestrol) caused clear cell adenocarcinoma (cancer of the cervix and vagina), birth defects, and other developmental abnormalities in children born to women who took the drug while pregnant. It also increased the risk of breast cancer, and increased the risk of cancer in children of mothers who took the drug. This drug was FDA approved.

Meridia (Sibutramine) caused increased cardiovascular and stroke risk. This drug was FDA approved.

Merital & Alival (Nomifensine) caused deaths due to immuno-hemolytic anemia. This drug was FDA approved.

Mylotarg (Gemtuzumab Ozogamicin) increased the risk of death and veno-occlusive disease (obstruction of veins). This drug was FDA approved.

Omniflox (Temafloxacin) caused deaths, severe low blood sugar, hemolytic anemia and other blood cell abnormalities, and kidney dysfunction (half of the cases required renal dialysis). This drug was FDA approved.

Pondimin (Fenfluramine) caused 30 percent of patients that had been prescribed the drug to have abnormal echocardiograms. This drug was FDA approved.

Posicor (Mibefradil) caused fatal interactions with at least twenty-five other drugs such as common antibiotics, antihistamines, and cancer drugs. This drug was FDA approved.

Propulsid (Cisapride) caused more than 270 cases of serious cardiac arrhythmias reported between July 1993 and May 1999, with seventy deaths. This drug was FDA approved.

Raptiva (Efalizumab) caused progressive multifocal leukoencephalopathy (PML). This is a rare and usually fatal disease that causes inflammation or progressive damage of the white matter of the brain. This drug was FDA approved.

Rezulin (Troglitazone) caused at least ninety liver failures and sixty-three deaths. This drug was FDA approved.

Selacryn (Tienilic acid) caused hepatitis, thirty-six deaths, and at least five hundred cases of severe liver and kidney damage. This drug was FDA approved.

Vioxx (Rofecoxib), a Merck drug, increased risk of heart attack and stroke. This drug was FDA approved.

Remember that all these FDA-approved drugs that injured or killed presumably were subject to placebo-controlled, double-blind studies. Vaccines are not even subject to this modest standard.

The bottom-line is that FDA approval does not mean a drug or vaccine is safe. The measles vaccine is FDA approved, but the nation's top scientists say "the evidence is inadequate" to determine if the vaccine causes other severe and life threatening diseases.

Sources:

"FDA-Approved Prescription Drugs Later Pulled from the Market." Prescription Drug Ads. Last modified February 18, 2020. https://prescriptiondrugs.procon.org/fda-approved -prescription-drugs-later-pulled-from-the-market/.

Stratton, Kathrine, Board on Population Health and Public Health Practice, and Committee to Review Adverse Effects of Vaccines. *Adverse Effects of Vaccines: Evidence and Causality.* Washington: National Academies Press, 2011.

The Measles Vaccine and All Vaccines Are Considered Legally "Unavoidably Unsafe"

Quick Version: Vaccine manufacturers do not improve their old vaccines like the sixty-year-old measles vaccine because they claim it is unavoidably unsafe. They say that there is nothing they can do to make the vaccine safer even though it was invented in the 1950s with primitive technologies.

Our government allows vaccine makers to treat their vaccines as "Unavoidably Unsafe."

In 1986, the US Congress decided that all vaccines are "unavoidably unsafe." What does that mean?

In the case of vaccines, it means that they are "incapable of being made safe for their intended and ordinary use." It means that there is nothing else that can be done to make the product safer without compromising its effectiveness.

This, of course, is a convenient assumption!

Which is safer—a car made today or one made in 1960?

Imagine, someone makes a product in 1960 (a car for example) with 1950s technology and sixty years later says that there is nothing they can do today to make the product any safer. Would you believe that?

Here's a very brief list of some of the improvements in car safety over the last sixty years that have saved tens of thousands of lives:

- Airbags
- Crumple zones
- Anti-lock braking
- Three point seat belt
- Lane departure warning systems
- Tire pressure monitoring
- High strength steel
- Back-up cameras
- Traction control
- Blind spot sensors

Why do we expect car manufacturers to improve their products, but we accept a drug company's position that their sixty-year-old product simply can't be made any safer?

Our government allows vaccine makers to use vaccines invented almost sixty years ago with no quality improvement whatsoever.

The measles vaccine was invented at the end of the 1950s and early 1960s. In the last sixty years the world of biotechnology has changed. Today we have: DNA sequencing, polymerase chain reaction (PCR), DNA cloning, gel electrophoresis, fluorescent in situ hybridization, genomics and proteomics, and many other new technologies.

Nonetheless, vaccine manufacturers like Merck tell the government that there is nothing they can do to make their vaccines safer without reducing their effectiveness.

This provides a great advantage to vaccine makers. They do not have to work hard or spend money to improve the safety of their vaccines and they don't have to pay when their vaccines hurt or kill a child.

Other products that have qualified as unavoidably unsafe include guns, cleaning compounds like commercial dry-cleaning solvent, industrial-strength bathroom cleaners, acetone, benzene, bleach, and dye. Congress decided that vaccines are in the same category.

Here's the legal language:

> No vaccine manufacturer shall be liable in a civil action for damages arising from a vaccine-related injury or death associated with the administration of a vaccine after October 1, 1988, if the injury or death resulted from side effects that were unavoidable even though the vaccine was properly prepared and was accompanied by proper directions and warnings.

Isn't it strange that products that are injected into our children are categorized the same as industrial strength bathroom cleaner, acetone, bleach, and other chemicals?

Shouldn't a different standard apply to any type of medical product given to healthy children? Shouldn't vaccine makers be required to improve the safety of products that were invented more than half a century ago?

Sources:

Language adopted by National Childhood Vaccine Injury Act, 42 USC 300aa-22: §402A of the Restatement of Torts (Second) (1963– 1964) (hereinafter Restatement), which provides that "unavoidably unsafe" products— i.e. , those that "in the present state of human knowledge, are quite incapable of being made safe for their intended and ordinary use . . . "See: Bruesewitz v. Wyeth, 562 U.S. (2011) https://injury.findlaw .com/product-liability/what-is-an- unavoidably-unsafe-product.html.

"BRUESEWITZ ET AL. v. WYETH LLC." *Supreme Court of the United States.* Last modified February 22, 2011. https://www.supremecourt.gov/opinions/10pdf/09–152. pdf.

Waxman, Henry A. *H.R.5546—99th Congress (1985–1986): National Childhood Vaccine Injury Act of 1986.* Washington, DC: Congress.org, October 18, 1986.

"What Is an Unavoidably Unsafe Product?" Findlaw. Last modified December 4, 2018. https://www.findlaw.com/injury/product-liability/what-is-an-unavoidably-unsafe -product.html.

SECRET #13

The Measles Vaccine Maker Admits There Are Over Seventy Potential Side Effects

> *Quick Version:* Merck admits in their vaccine insert that over seventy side effects are possible with the measles vaccine.

Merck makes the measles vaccines. By Merck's own admission, there are a number of potential side effects associated with this vaccine. The following is a list from the manufacturer's insert, the FDA required document that describes what is known about a particular vaccine or medication.

See what Merck says about what diseases and side effects are associated with the measles vaccine.

Since vaccines are given by doctors or nurses, and not administered by ourselves like other medications we might take, we never see this insert. A glossary is included at the back of the book so you can see what some of these terms mean.

Here is what Merck says:

The following adverse reactions are listed in decreasing order of severity, without regard to causality, within each body system category and have been reported during clinical trials, with use of the marketed vaccine, or with use of monovalent or bivalent vaccine containing measles, mumps, or rubella:

Panniculitis; atypical measles; fever; syncope; headache; dizziness; malaise; irritability; Vasculitis, Pancreatitis; diarrhea; vomiting; parotitis; nausea; Diabetes mellitus; Hemic and Lymphatic System; Thrombocytopenia purpura; regional lymphadenopathy; leukocytosis; Anaphylaxis and anaphylactoid reactions have been reported as well as related phenomena such as angioneurotic edema bronchial spasm; Arthritis; arthralgia; myalgia; Encephalitis; encephalopathy; measles inclusion body encephalitis; subacute sclerosing panencephalitis (SSPE); Guillain-Barré Syndrome (GBS); acute disseminated encephalomyelitis (ADEM); transverse myelitis; febrile convulsions; afebrile convulsions or seizures; ataxia; polyneuritis; polyneuropathy; ocular palsies; paresthesia. In severely immunocompromised individuals who have been inadvertently vaccinated with measles containing vaccine; measles inclusion body encephalitis, pneumonitis, and fatal outcome as a direct consequence of disseminated measles vaccine virus infection have been reported. Pneumonia; pneumonitis; sore throat; cough; rhinitis.; Stevens-Johnson syndrome; erythema multiforme; urticaria; rash; measles-like rash; pruritis; Local reactions including burning/stinging at injection site; wheal and flare; redness (erythema); swelling; induration; tenderness; vesiculation at injection site; Henoch-Schönlein purpura; acute hemorrhagic edema of infancy; Nerve deafness; otitis media; Retinitis; optic neuritis; papillitis; retrobulbar neuritis; conjunctivitis; Urogenital System; Epididymitis; orchitis.

So, in the few studies that have been done, Merck has already concluded that these side effects are possible. This shouldn't be surprising. Do you remember the studies done in the early 1960s described in a previous chapter? In those studies, scientists had already discovered that the measles vaccine was associated with bronchitis, seizures, cyanosis, and aberrations of encephalographic patterns.

But, despite all these side effects, Merck says they cannot make their sixty-year-old vaccine any safer.

Source:

"Highlights of Prescribing Information M-M-R II." *Merck.* Accessed February 11, 2021. https://www.merck.com/product/usa/pi_circulars/m/mmr_ii/mmr_ii_pi.pdf.

According to the CDC and FDA, There Are 89,032 Cases of Injury Associated with the Measles Vaccine, or Is It a Hundred Times as Much?

> *Quick Version:* A system called VAERS has collected reports of tens of thousands of injuries and many deaths from the measles vaccine, but this number may be under-reported by almost a hundred times.

The VAERS database collects information on vaccine side effects.

The CDC and FDA collect statistics about injuries and deaths associated with vaccines. The database they use is called VAERS (Vaccine Adverse Event Reporting System).

According to the CDC:

> VAERS is a national vaccine safety surveillance program run by CDC and the Food and Drug Administration (FDA).
>
> VAERS serves as an early warning system to detect possible safety issues with U.S. vaccines by collecting information about adverse events that occur after vaccination.

A search for injuries associated with the measles vaccine from March 1990 to October 2019 produced 89,032 case reports.

The data is not perfect because anyone can report an injury or death and there is no reporting requirement. The CDC and the FDA have decided, however, that the information is reliable. Why? Because they rely on it themselves for their decision making. If they wanted to improve it, they would.

They could, for example, require that all vaccine-related injuries be reported within forty-eight hours or a fine would be paid. This would help ensure that this important data is being collected by healthcare providers. But, strangely, the health authorities don't do this.

They could also require that all serious injuries or deaths associated with a vaccine have follow-up by experienced clinicians and researchers. But this isn't done either.

Isn't this data important? Why doesn't the FDA or CDC care to get a comprehensive report when a child is injured or killed after vaccination?

This lack of interest demonstrates the real motivation of our health authorities. Despite all the rhetoric to the contrary, the health authorities stick their head in the sand when it comes to vaccine injury. They really don't want to know the truth. They would much rather put their energy and efforts in forcing us all to vaccinate. They are much less interested in knowing what happens when we do.

The Harvard Pilgrim Health Care Report suggests that there could be a hundred times more vaccine injuries.

A report that received zero media attention was published by Harvard Pilgrim Health Care, Inc. It states that: "Fewer than 1% of vaccine adverse events are reported."

This finding was submitted to the Agency for Healthcare Research and Quality (AHRQ) at the US Department of Health and Human Services (HHS).

This suggests that there may be a hundred times as many injuries and deaths related to vaccines than what is reported in VAERS

The reasons for the low reporting rate (approximately 1 percent) were investigated and found to include:

> [A] lack of clinician awareness, uncertainty about when and what to report, as well as the burdens of reporting: reporting is not part of clinicians' usual workflow, takes time, and is duplicative.

This illuminating report concluded that:

> Low reporting rates preclude or slow the identification of 'problem' drugs and
> vaccines that endanger public health. New surveillance methods for drug and
> vaccine adverse effects are needed.

The study concluded that it's *not* hard to improve the system because of "proactive, spontaneous, automated adverse event reporting imbedded within electronic health records."

In other words, with electronic medical records used by most health care organizations today, there is already built-in functionality to capture this vaccine injury data.

Guess what happened when the researchers asked the CDC to run a test to see how easy it would be to capture data about vaccine side effects? The CDC was not interested.

The report says:

> Unfortunately, there was never an opportunity to perform system perfor-
> mance assessments because the necessary CDC contacts were no longer avail-
> able and the CDC consultants responsible for receiving data were no longer
> responsive to our multiple requests to proceed with testing and evaluation.

Think about that.

A study funded by the US government finds that the current system to collect vaccine side effects (VAERS) captures only about 1 percent of vaccine injuries. The study then suggests how to make VAERS much more accurate, but the CDC was not "responsive to our multiple requests . . . " They were not interested.

This speaks volumes.

In summary:

- There is a national reporting system designed to identify injuries from vaccines.
- A reputable study found that this system is inadequate and captures only about 1 percent of the vaccine related injuries.
- The study suggested that "New surveillance methods for drug and vaccine adverse effects are needed." Suggestions were made how to improve this system.

- The FDA and CDC have responded with crickets. Silence. Is it possible that these agencies don't want a better system to collect and report on vaccine injuries, especially if the numbers are one hundred times what is being collected now?
- Even though the present system is woefully inadequate, it has still collected 89,032 reports of injury associated with the measles vaccine.
- If the system collected all adverse events it could be one hundred times as much or 8,903,200. That's a frightening possibility.

Bottom line: there are already tens of thousands of vaccine-related deaths and injuries reported and the government is not interested in creating an accurate system to collect all of them. Think about this.

Source:

Lazarus, Ross. "Electronic Support for Public Health–Vaccine Adverse Event Reporting System (ESP:VAERS)." *AHRQ Digital Healthcare Research.* Last modified December 1, 2007. https://healthit.ahrq.gov/sites/default/files/docs/publication/r18hs017045-lazarus -final-report-2011.pdf.

SECRET #15

The US Government Has Children Injured by the Measles Vaccine in Their Database

> *Quick Version:* Here are examples of side effects that occurred immediately after the measles vaccine was given. They include: difficulty walking, abnormal behavior, anaphylactic reactions, bronchitis, convulsions, seizures, and neurological side effects.

Despite the fact that injuries and deaths may be under-reported by up to a hundred times, there are already thousands of injuries of hundreds of different side effects reported that immediately follow the administration of the measles vaccine. These reports can be found in the VAERS system available here: https://wonder.cdc.gov/vaers.html.

It can be difficult to determine which injuries were caused by the measles vaccine or another vaccine that the child received at the same time.

VAERS says that just because an injury happens after a vaccine does not mean that the vaccine caused it. That's a convenient position to take.

Let's think about that. The government mandates vaccines. Children are injured. Some (maybe one percent) of the injuries are reported to a national database. The government does not facilitate any independent review to determine if the vaccine was responsible. Because there is no follow-up, the government can deny the connection.

Did the vaccines cause these injuries? Sadly, you have to draw your own conclusions. It shouldn't be this way.

The examples that follow are just a tiny fraction of the side effects associated with the measles vaccine and other vaccines administered the same day. These examples were selected from VAERS because the side effect took place within hours of the vaccination. You can see that the "Days of Onset" for each case is "0," which means the side effect took place on the same day that the vaccine was given.

Also, the age of each child is reported in years instead of months. So for example 1.33 would be sixteen months (1.0 = 12 months and .33 is 1/3rd of a year or 4 months; 12 + 4 = 16 months).

983 cases of difficulty walking associated with measles vaccine

Example from VAERS:

Patient Age	1.33	Sex	Female
State / Territory	Oregon	Date Report Completed	2017-07-11
Date Vaccinated	2017-06-26	Date Report Received	2017-07-11
Date of Onset	2017-06-26	Days to Onset	0

Vaccines Given:
MMR II
PREVNAR 13
VARIVAX

Extreme leg pain (wouldn't walk, didn't want held, rolled around on floor in agony). Screamed and cried for 6 hours after shot. Fever started off and on for almost 2 weeks (up to 103 but don't know what the spike that caused the seizure was). Febrile seizure (7/5/17). Patient's paralysis RT side (7/5/17). Helicopter (7/5/17). Rash (7/7/17–7/9/17). Tylenol and Ibuprofen wasn't controlling fever. Loss of appetite. Had to give her bottle to feed her. Stopped breathing (blue lips, limp body) (7/5/17). Had to do CPR (7/5/17). Irritable, moody, very unhappy, sad, cried for no apparent reason through the day and night, extremely tired but wouldn't sleep for 2 weeks after shots. Diarrhea for a week after shots. Throwing up 1–3 times per day after shots until 7/3/17. Wouldn't even take a bath without a screaming tantrum after vaccines until 7/10/17. Seizure caused her to fall off a chair and hit her head. VAERS ID # 703369–1

541 cases of abnormal behavior associated with measles vaccine
Examples from VAERS:

Patient Age	2.00	Sex	Female
State / Territory	California	Date Report Completed	2019-08-17
Date Vaccinated	2019-03-25	Date Report Received	2019-08-17
Date of Onset	2019-03-25	Days to Onset	0

Vaccines Given:
MMR II

Patient had marked change in behavior and became very irritable, aggressive and did not stop crying for 3 days, she stopped talking normally for two weeks. Patient could not sleep and was very fussy. She had a fever 5 days after the vaccine of 106.7 and had a full body rash. She did not have much appetite. Patient had marked emotional, verbal, and physical regressions. VAERS ID # 829176–1

Patient Age	1.00	Sex	Male
State / Territory	Iowa	Date Report Completed	2019-08-15
Date Vaccinated	2019-08-07	Date Report Received	2019-08-15
Date of Onset	2019-08-07	Days to Onset	0

Vaccines Given:
MMRII
HIB

My son doctor appointment was on the 7th at 4:00pm he was a healthy happy fine baby a baby who love to talk and do sign language since 6 months old. After receiving the vaccines 1 half an hour after my son was a whole new person. He dont talk no more or smile he hate me will not come to me or look at not even respond to his name no more. VAERS ID # 828750–1

Patient Age	1.00	Sex	Female
State / Territory	Oklahoma	Date Report Completed	2019-06-17
Date Vaccinated	2019-05-02	Date Report Received	2019-06-17
Date of Onset	2019-05-02	Days to Onset	0

Vaccines Given:
MMR II
DTAP
HIB
HEPATITIS A
PREVNAR
VARIVAX

Prolonged irritability after 12 month vaccines/mmr. Continuing to this day. Patient does not sleep well at night, wakes up multiple times during the night screaming. Does not act the same as before shots. VAERS ID # 819136–1

Patient Age	4.00	Sex	Female
State / Territory	Unknown	Date Report Completed	2019-05-10
Date Vaccinated	2016-01-01	Date Report Received	2019-05-10
Date of Onset	2016-01-01	Days to Onset	0

Vaccines Given:
MMR II
DTAP
POLIOVIRUS VACCINE INACTIVATED
(VARIVAX)

Shortly after vaccination, my daughter's behavior spiraled out of control. She began having rage episodes and sensory meltdowns multiple times a day that lasted up to an hour per episode. She began stimming, wringing her hands, chewing on her fingers, scooting along the floor and needing constant physical stimulation. She was unable to sit still for more than 1–2 minutes at a time. She became extremely inflexible and rigid in personality, unable to handle even the slightest change in schedule. She began mixing up her letters. She was eventually diagnosed with PANDAS and multiple chemical sensitivity one year after her 5 year shots (as documented above), during which time we were actively trying to pinpoint what was wrong with her. It is obvious that it began with this vaccination. VAERS ID # 813467–1

264 cases of anaphylactic reaction and anaphylactic shock associated with measles vaccine

Examples from VAERS:

Patient Age	2.00	Sex	Male
State / Territory	Michigan	Date Report Completed	2019-03-28
Date Vaccinated	2019-03-28	Date Report Received	2019-03-28
Date of Onset	2019-03-28	Days to Onset	0

Vaccines Given:
MMRII

10 minutes after receiving the MMR vaccine he walked to the front desk with his mother, he sneezed 5 times with mucus coming out of his nose, he then had a gagging cough and threw up once, he then continued to have dry heaves and he looked puffy around his eyes, he then started to become lethargic alternating with combative, his O2 dropped to 88%, he face became puffier, we give him AUVI-Q 0.15 in the left thigh and called 911, oxygen was started as his O2 was 100%, he was lethargic off and on, but his HR, RR and BP remained stable, he never developed a rash, he was taken by ambulance to the local hospital and given IV SOLU-MEDROL and diphenhydramine, after about 4 hours he became much more active, less swollen and he was drinking, he was discharged home from the ER with a EPIPEN, he will follow up with an allergist; the clinical picture was consistent with anaphylaxis. VAERS ID # 807550–1

Patient Age	5.00	Sex	Male
State / Territory	Illinois	Date Report Completed	2008-08-25
Date Vaccinated	2008-08-20	Date Report Received	2008-08-25
Date of Onset	2008-08-20	Days to Onset	0

Vaccines:
MMR II
DTAP (INFANRIX)
POLIOVIRUS VACCINE INACTIVATED
VARICELLA (VARIVAX

5 yr old who developed moderate to severe respiratory distress, altered mental status and urticaria within minutes of receiving MMM2,Varicella2, DTaP5, IPV4. BP 63/32. Received subq epi x 3 and O2 when pulse ox dropped to upper 80s. EMS took patient to ER where he was observed for 6 hours and was stable. Mom reported that he had a previous "allergic reaction" after receiving Hep A and Influenza vaccine (hives) on 11/20/06. Was referred to A&I multiple times but did not go 9/4/08-ED records received for DOS 8/20/08-Assessment:Allergic reaction. Presented to ED with anaphylactic reaction of facial swelling, wheezing shortness of breath and hypotension immediately after receiving Varicella." VAERS ID # 323207–1

294 cases of loss of ability to understand or express speech, caused by brain damage (aphasia)

Examples from VAERS:

Patient Age	1.00	Sex	Female
State / Territory	Massachusetts	Date Report Completed	2019-04-30
Date Vaccinated	2018-02-02	Date Report Received	2019-04-30
Date of Onset	2018-02-02	Days to Onset	0

Vaccines Given:
MMRII
VARICELLA (VARIVAX)

Loss of eye contact, loss of speech, Autism. Caused by two doses of MMR vaccine on 02/02/2018 & 08/03/2018. VAERS ID # 812060–1

Patient Age	1.00	Sex	Female
State / Territory	Mississippi	Date Report Completed	2019-02-06
Date Vaccinated	2016-10-27	Date Report Received	2019-02-06
Date of Onset	2016-10-27	Days to Onset	0

Vaccines Given:
MMRII
VARICELLA (VARIVAX)
Prevnar 13

Legs began to swell with large knots on both legs at the injection site, lasting a few weeks. She began running a fever which lasted about a week. She would no longer speak, hold eye contact, clap, wave, give hugs and kisses, play with her siblings, etc. She would just sit and stare or play alone. She would no longer respond to her name or any sounds to get her attention. She was tested on August 2017 and in less than 10 minutes they were able to diagnose her with autism spectrum disorder. She has been in therapy over a year now and is still non-verbal. VAERS ID # 800384–1

Patient Age	1.00	Sex	Male
State / Territory	Unknown	Date Report Completed	2019-02-05
Date Vaccinated	2016-02-08	Date Report Received	2019-02-05
Date of Onset	2016-02-08	Days to Onset	0

Vaccines Given:
MMRII

Fever, rash covering body, trouble waking him up, swollen and painful injection site. Within a week he was unable to walk and was in therapy until age 2. He did not walk again until 21 months. He stopped eating solid and was also in therapy to help as well with feeding. He stopped talking and was diagnosed with autism a month later. VAERS ID # 800239–1

Patient Age	1.25	Sex	Male
State / Territory	New Jersey	Date Report Completed	2018-05-04
Date Vaccinated	2014-01-09	Date Report Received	2018-05-04
Date of Onset	2014-01-09	Days to Onset	0

Vaccines Given:
HIB
MMR II
VARICELLA (VARIVAX)

Shortly after he developed high fevers and screaming. Followed by full body rash. My son has never been the same. He is currently in weekly private OT, speech, and PT. He is in a multi disability kindergarten. There he receives the same services. He mentally acts like a 3 year old. Is unable to function like other children his own age. He is given life skills and sign language classes. Days after the vaccination he lost all capability for learned things. He was completely unintelligible and became angry and enraged over every little thing. VAERS ID # 747199–1

145 cases of bronchitis and bronchial-related side effects

Examples from VAERS:

Patient Age	4.00	Sex	Female
State / Territory	Connecticut	Date Report Completed	2014-08-13
Date Vaccinated	2014-08-12	Date Report Received	2014-08-19
Date of Onset	2014-08-12	Days to Onset	0

Vaccines Given:
MMR II
DTAP + IPV
VARICELLA (VARIVAX)

Less than 40 min after receiving MMR #2 and VZV #2, and KINRIX (DTaP #4, IPV-polio #4) child developed streaming eye, clear rhinorrhea, coughing, "tight" chest—bronchial constriction with wheezing, tongue slightly thickened and thickened voice. Initially given diphenhydramine 25 mg po for watery eyes, then given EPI PEN 0.15 mg for resp. symptoms. VAERS ID # 540328–1

Patient Age	5.00	Sex	Male
State / Territory	Pennsylvania	Date Report Completed	2013-03-20
Date Vaccinated	2013-03-18	Date Report Received	2013-03-20
Date of Onset	2013-03-18	Days to Onset	0

Vaccines Given:
MMR II
DTAP
POLIOVIRUS VACCINE INACTIVATED
VARICELLA (VARIVAX)

Child was given immunizations [dtap, ipv, mmr, varivax]. few minutes after administered vaccines, developed rhinorrhea, salivation, lacrimation, cough and bronchiospasm. immediately was treated with albuterol via neb and epinephrine 1:10,000 sc. VAERS ID # 487347–1

2,955 cases of convulsions-related side effects

Examples from VAERS:

Patient Age	1.05	Sex	Male
State / Territory	Washington	Date Report Completed	2015-02-20
Date Vaccinated	2011-03-25	Date Report Received	2015-02-20
Date of Onset	2011-03-25	Days to Onset	0

Vaccines Given:
MMRII

Screaming that could not be stopped. Shaking. Seizure without fever, and hitting his own head for more than three weeks after shot on 3/2011. He also has had years of diagnosed but unresolved swelling and pain. However night time is often the worst. He will cry out and wake sobbing in pain. This has been almost three years of night time pain in which we have sought help at several of the specialists at hospital. Loss of muscle tone and now physical delays. VAERS ID # 566115–1

Patient Age	13.00	Sex	Female
State / Territory	Illinois	Date Report Completed	2014-08-08
Date Vaccinated	2014-07-30	Date Report Received	2014-08-08
Date of Onset	2014-07-30	Days to Onset	0

Vaccines Given:
MMR II
DTP + IPV
HEP B (ENGERIX-B)
HPV (GARDASIL)

Immediately following immunizations we were walked out of room by nurse to the front desk to schedule a follow up appointment. Within 1–2 minutes of standing at the desk I heard a loud bang! I turned around to see what it was and my daughter was laying flat on her back, her eyes were rolling and she was seizuring. My daughter fainted while standing, fell back and slammed her head off the tile floor. She was unresponsive and very confused. 911 was called and she was taken by ambulance to the hospital. My daughter did indeed have a concussion. Since the fall she is experiencing severe headaches and neck pain started today. VAERS ID # 539264–1

Patient Age	1.04	Sex	Male
State / Territory	Unknown	Date Report Completed	2013-12-09
Date Vaccinated	2013-12-06	Date Report Received	2013-12-09
Date of Onset	2013-12-06	Days to Onset	0

Vaccines Given:
MMR II
FLUZONE
VARIVAX

Developed fever to 100.6. Proceeded to have 5 generalized tonic clonic seizures each lasting approximately 10 minutes with 10 minutes of drowsiness in between each episode, not returning to baseline. Presented to emergency room after 5th seizure where temperature was 103 and had additional generalized tonic clonic seizure lasting 40 seconds. VAERS ID # 516025-1

342 cases of seizure-related side effects
Examples from VAERS:

Patient Age	17.00	Sex	Male
State / Territory	Massachusetts	Date Report Completed	2017-11-14
Date Vaccinated	2017-11-14	Date Report Received	2017-11-14
Date of Onset	2017-11-14	Days to Onset	0

Vaccines Given:
MMR II
HEP A (VAQTA)
HEPATITIS B
HPV (GARDASIL 9)
FLUZONE QUADRIVALENT
POLIO VIRUS, INACT. (IPOL)
TETANUS AND DIPHTHERIA TOXOIDS

After receiving the patient's last vaccine, the RN reported that the patient started to faint and she had to catch him. It was unclear if he had a seizure or a vasovagal episode for us. The patient became awake alert and oriented after about 10–15 secs. While, the patient was describing what he felt which was a sensation of black from his eyes and fainting and then waking up. The patient had a second episode that was directly witnessed by this provider. The patient specifically started arching his back and leaned toward his left. His biceps contracted and he kept them close to his chest. His legs started having more classic tonic clonic movements. The patient was laid down. This episode lasted approximately 15 seconds and then patient was awake/alert afterwards. AAOx3. The timing between the episodes was approximately 5 minutes. Decision was made to transfer to ER downstairs. Unclear if first episode was a seizure or a vasovagal episode but second episode was more classic convulsion. Probably vasovagal as he has no prior history of seizures. Of note, two episodes is atypical. The episodes would be best described as LOC with myoclonus of lower extremities. VAERS ID # 726620–1

Patient Age	1.00	Sex	Male
State / Territory	Wyoming	Date Report Completed	2019-09-02
Date Vaccinated	2018-08-14	Date Report Received	2019-09-02
Date of Onset	2018-08-14	Days to Onset	0

Vaccines Given:
MMRII
Prevnar 13

2 seizures less than 8 hours after. My son seized for about 2 minutes, stopped breathing, and fever spiked to 102 degrees. 2nd seizure less than 45 minutes after 1st while in hospital. Again stopped breathing, full body seizure, fever at 102 degrees. VAERS ID # 831414–1

159 cases of neurologic or mental related side effects
Example from VAERS:

Patient Age	4.00	Sex	Male
State / Territory	Nevada	Date Report Completed	2018-02-26
Date Vaccinated	2018-02-24	Date Report Received	2018-02-26
Date of Onset	2018-02-24	Days to Onset	0

Vaccines Given:
MMRV (PROQUAD)
DTAP + IPV (QUADRACEL)
FLUARIX QUADRIVALENT

Following administration of Vaccines, patient experienced decreased BP (76/47), syncope, vomited x2 (clear bilious) and blanching of skin around eyes and to fingers of hands bilaterally. Cheeks and dorsal surface of hands became erythematous. Patient placed in supine position with LE raised. BP retaken 86/50 HR 90 O2 Sat=98%—no noted respiratory distress and lungs clear but mentation decreased, "dazed" appearance with blanching to periorbital area and fingers bilaterally, and erythema to cheeks and dorsal surface of hands remaining. BP retaken after lying supine for 10min—92/47 but when taken to standing BP decreased to 76/48 HR 90 O2Sat=100%. Mentation remained decreased and "dazed" appearance persisted. 911 called after 40 minutes of efforts to stabilize unsuccessful—EPIPEN and Diphenhydramine administered. Dr. at ED called and report given. Transported to ER at ~1045hrs via EMS. VAERS ID # 738934–1

Patient Age	4.00	Sex	Male
State / Territory	California	Date Report Completed	2016-09-23
Date Vaccinated	2016-09-21	Date Report Received	2016-09-23
Date of Onset	2016-09-21	Days to Onset	0

Vaccines Given:
MMR II
DTAP
POLIOVIRUS VACCINE INACTIVATED
VARIVAX

Around 12:54 pm, the father frantically carried him into the clinic. Father passed the patient to me and I carried him into room 2 onto the table. He was slightly responsive but had foaming saliva around the mouth, and when his mother asked him to repeat his sister's name he repeated it. He was not in respiratory distress. His father denied any seizures, but said he was almost passing out. Blood glucose was checked and it was 118 by MA. The patient responded to the fingerprick and cried normally. Pulse ox with portable finger oximetry was at 65–70% but that was rechecked with a child pulse oximetry which confirmed the low oxygen level. MA, brought the oxygen. Oxygen 10 L face mask was given. Because the patient did not like the face mask, we sat him up in his father's lap. We removed the oxygen and his oxygen level was more than 95%. However, then he lost consciousness for less than 10 seconds. No seizure activity. Once we moved him back to the table and put on oxygen he woke up again. That is when we noticed the hive over 5–7 cm oval next to the left leg injection site. Because he had a change in mental status, I decided to give epinephrine 0.15 mg SQ. The injection was given by RN at 1:03 pm. 911 was called by MA. After the injection, the patient's mental status returned to normal and he started talking. He face and body became flushed and there was mild swelling in the extremities. VAERS ID # 654992–1

The VAERS Database Says There Are 275 Deaths Associated with the Measles Vaccine

> *Quick Version:* Here are some examples of deaths of children immediately following the measles vaccine. These examples come from the government's own database.

A search of the VAERS (the Vaccine Adverse Event Reporting System) database created by the FDA and CDC reveals 275 deaths associated with the measles vaccine from 1990 until October 2019.

If we look at all vaccines (not just measles or vaccines administered with measles), there were 4,690 vaccine related deaths reported in VAERS from March 1990 to October 2019. This averages to approximately 156 deaths per year or three deaths a week. The accuracy of this number is in question for two reasons:

1. As discussed previously, many instances of injury or death go unreported. The Harvard Pilgrim Health Care study suggested that 99 percent of vaccine injuries and deaths may go unreported.
2. There may be deaths and injuries that are totally unrelated to vaccines. However this can be partly addressed by focusing on those

injuries and deaths that took place on the same day of vaccination
or soon thereafter.

Here are some recent examples. The names of the children were redacted by
VAERS but you can go to https://wonder.cdc.gov/controller/datarequest/
D8 and use their VAERS ID number to learn more about these cases.

Did the measles vaccine cause the death or is the vaccine merely associated with the death? The government has the responsibility of answering
that question. But, they don't.

The government has the responsibility of guaranteeing the safety of our
children when they mandate a mass vaccination program. That means that
we shouldn't be guessing when it comes to vaccine injuries and deaths. The
government, however, is not doing its job.

None of these death cases are being followed up on by government
funded independent research. So there is no clarity. All we have are reports
of children dying after getting vaccines. Because the government neglects
its duty when it comes to ensuring the safety of vaccines, parents are left to
draw their own conclusions.

Did the vaccines cause these deaths? You can decide. It shouldn't be this
way.

Deaths associated with the measles vaccine

- In July 2014, a boy, age 1.5 years old, received MMR and DTAP
  vaccines. According to the report he was "found dead the next
  morning by family member." (VAERS ID # 537090)
- In October 2014, a boy age 1.3 years old, received MMR and two
  other vaccines. According to the report, "Was at daycare—had
  taken nap/aroused fussy—consoled/but then noted not breathing
  911—to ER/dx with cardiac arrest." (VAERS ID # 546331)
- On August 31, 2015, a twenty-one-month-old girl was administered
  eleven vaccines at once (MMRV, DPT, polio, Hib, Hep A, and
  Prevnar). The report states, "Dr received phone call from ER at
  9:46 am. Dr at ER states child presented to the ER via personal
  automobile. Mom states patient was gagging all night. Child was
  in cardiac arrest when she arrived in the ER at 8:16 am. Pediatric
  advanced life support protocol active for one hour using every
  cardiac drug per Dr." (VAERS ID # 592715–1)
- On October 5, 2015 a thirteen-month-old boy was vaccinated with
  HIB, Hep A, Hep B, MMRV, and Prevnar. According to the

report, "Pt. received physical and shots on 10/5/15 and passed 10/8/15. Per family he had a fever x2d, no fever on 3rd day or 4th day and then died suddenly on 10/8/15. The coroner's preliminary slides show viral myocarditis and interstitial pneumonia." (VAERS ID # 603163–1)

- On October 1, 2018, a boy, age twelve months, received Hep A, Flu and MMRV (Proquad) vaccines. According to the report, "Received vaccines 10/1/18. Patient developed fever on 10/3/18. Temperature max 101. Mom gave Motrin before bed that night. Checked on him before she went to bed and his temp was 99. Next morning he was found dead in crib. No blankets, stuffed animals in crib. Autopsy found no signs of trauma." (VAERS ID # 776286–1)

Deaths associated with other vaccines

- On July 1, 2019, a six-week-old girl was vaccinated with HBV (Engerix B). The next day, the girl died. The reported cause of death was unknown. The reporter considered the unknown cause of death to be related to Engerix B. The patient was not known to have any pre-existing health conditions. (VAERS ID # 823271–1)
- On November 29, 2018, a three-month-old boy was vaccinated with DTAP + IPV + HIB and HEP B and ROTAVIRUS VACCINE. He died the same day. The report states, "Vaccines were given in the morning of 11/29/2018. Pt was lethargic and was refusing to eat. On 11/29/2018 patient passed away in his crib, sleeping around 10:45am." (VAERS ID # 819187–1)
- On May 21, 2019, a six-month-old girl was vaccinated with DTAP + HEP B + IPV (PEDIARIX), HIB (HIBERIX), PREVNAR13 and ROTAVIRUS (ROTATEQ). She died the next day. The report states, "Child tolerated administration of vaccines well. Child left office with family after appointment. Our office was notified the following day (5/22/2019) that child died overnight. She was found unresponsive. She was taken to an Emergency Department. Resuscitation was attempted." (VAERS ID # 815652–1)
- On April 18, 2019, a three-month-old boy was vaccinated with HIB, PREVNAR13, and ROTAVIRUS (ROTATEQ). He died within two days. The report states, "The child had seen his pediatric provider 2 days prior to presentation to emergency department and had immunizations. The day prior to presentation, the child was seen for loose stools. He was described as nontoxic in appearance

and feeding okay with normal urinary output. Presented to the emergency department via ambulance with CPR in progress after suffering an in-home arrest. Resuscitation was terminated after patient's arrival and patient was pronounced as deceased." (VAERS ID # 810943–1)

• On February 7, 2019, a ten-month-old girl received DTAP, HIB (HIBERIX), HEP B (ENGERIX-B), INFLUENZA VIRUS VACCINE, PNEUMOCOCCAL VACCINE, and POLIOVIRUS VACCINE INACTIVATED. She died less than a month later. The report stated, "4 hours after my daughter got her shots her fever wouldn't go down. At 11pm that night she began to have complex febrile seizures. She had 3 that night. After coming home she wasn't the same. One of her eyes didn't open as much as it should have and she was slower. Almost a month later she died in her sleep because of another febrile seizure." (VAERS ID # 806371–1)

• On February 27, 2019, a six-month-old girl was vaccinated with DIPHTHERIA AND TETANUS TOXOIDS AND ACELLULAR PERTUSSIS VACCINE, HEPATITIS B, and INACTIVATED POLIOVIRUS VACCINE. She died March 1, 2019 (two days after vaccination). The report stated, "Patient seen for routine well child exam 2-27- 2019 with no abnormal findings." (VAERS ID # 804153–1)

• On February 12, 2019, a four-month-old boy was vaccinated with DIPHTHERIA AND TETANUS TOXOIDS AND ACELLULAR PERTUSSIS VACCINE, HEPATITIS B, INACTIVATED POLIOVIRUS VACCINE, HAEMOPHILUS B CONJUGATE VACCINE, PREVNAR, and ROTAVIRUS VACCINE. He died on February 13, 2019 (the next day). The report states, "Child was in office on February 12th for his four month well visit where he received four vaccines, PEDIARIX, PREVNAR, Hib and Rotavirus. While in office he was alert and happy, smiling, and active. Per mother he was also alert and happy at home until he was put down for tummy time by the father around 5pm. Previously, child was fed by dad and laid down on his stomach on the parent's bed for tummy time, per dad. Father stepped out of the room to the living room and fell asleep. Next time child was checked on was once mom got home from work at 10:30pm and mom noticed he was blue but still breathing. Mother called 911 and once paramedics arrived he seemed to be having a seizure and was given medication.

Child was rushed to hospital where he was put in NICU, and doctors stated there was nothing else that could be done. Child was disconnected the morning of Feb. 13th, 2019." (VAERS ID # 801369–1)

- On January 10, 2019, a nine-month-old boy was vaccinated with INFLUENZA VIRUS VACCINE, QUADRIVALENT (INJECTED). He died the same day. The report states, "Per ER report, patient received his Flu vaccine at his 9 month well check. Parents took him home. He was fed and then went down for a nap. One to two hours later mother reports finding the child cyanotic and unresponsive. EMS called and found him in asystole. EMS attempted resuscitation en route to ER by placing IO and attempted intubation. He was given epinephrine x Q CPR continued in the ED. Warmed to 35.4, intubated with 3.5 cuffed ET and received multiple rounds of epinephrine. Chest compressions were recorded for 18 minutes. Time of death was called at 18:14." (VAERS ID # 795791–1)
- On November 9, 2018, an eight-month-old boy was administered INFLUENZA (FLULAVAL QUADRIVALENT). He died on November 12, 2018 (3 days after he was vaccinated). The report states, "Several hours after vaccine was administered, Mother found child unresponsive, not breathing and without pulse. Child taken to medical center via EMT. Other circumstances, Child was sleeping and mother stated found child blue with a teddy bear." (VAERS ID # 785708–1)

Many Doctors Admit in Their Publications that the Measles Vaccine Can Cause Dangerous Side Effects

Quick Version: There are many thousands of articles in the peer reviewed literature (journal articles written by researchers and doctors) that describe injuries following the measles vaccine.

The government has decided that it's not important to have a formal follow-up program to determine what caused the deaths reported in their VAERS database. For the vaccine industry it's better to stick its head in the sand and tell the public that deaths and serious injuries associated with vaccines have not been proven.

Therefore, there is no program designed for medical follow-up when a child is suspected to have been killed or severely injured from a vaccine. This is a strange omission. If the government cared about the safety of its vaccine program it should find out why children are being injured or dying.

Doctors write up reports of some children injured or killed by vaccines.

However, medical doctors and scientists have known for years that the measles vaccine can cause or is associated with acute, chronic, and life-threatening side effects in some children. There are thousands of medical articles about this. These articles are often case studies. So while there is no formal program to conduct medical follow-up on vaccine injuries, some independent doctors and scientists have been concerned enough that they have conducted their own investigations.

These investigations are very revealing.

The following articles have all been published in peer-reviewed medical journals. To be clear, the conclusions in some papers are that the vaccine caused the injury. Other papers suggest it didn't, or more research is needed. The bigger point is that these vaccine-related injuries are of enough concern that some scientists are studying them on their own.

All of the following articles can be found in the National Institute of Health's PubMed database. PubMed comprises more than twenty-nine million citations for biomedical literature from MEDLINE, life science journals, and online books. You can access it here: https://www.ncbi.nlm.nih.gov/pubmed/.

Some of the articles are written by doctors at universities and teaching hospitals here in the United States and some are from other countries. Remember that nearly all the vaccines they are discussing are the same ones available in the United States and are made by the same companies.

This is a very small sampling of the titles of some of the articles. You can read the abstracts (summaries) of the actual articles by entering these titles into a Pubmed search:

Clinical and laboratory features of aseptic meningitis associated with measles-mumps-rubella vaccine
Rev Panam Salud Publica. 2002

Haemophagocytic lymphohistiocytosis following measles vaccination.
Eur J Pediatr. 2002

Gianotti-Crosti syndrome after measles, mumps and rubella vaccination.
Br J Dermatol. 1998

Measles, mumps, rubella vaccine induced subacute sclerosing panencephalitis.
J Indian Med Assoc. 1997

Optic neuritis following measles/rubella vaccination in two 13- year-old children.
Br J Ophthalmol. 1996

Arthritis after mumps and measles vaccination.
Arch Dis Child. 1995

Hearing loss following measles vaccination.
J Infect. 1995

Gait disturbances after measles, mumps, and rubella vaccine.
Lancet. 1995

Neurologic disorders after vaccination against measles and mumps.
Cesk Pediatr. 1992

Joint and limb symptoms in children after immunisation with measles, mumps, and rubella vaccine.
BMJ. 1992

Postvaccinal parkinsonism.
Mov Disord. 1992

Risk of seizures after measles-mumps-rubella immunization.
Pediatrics. 1991

Bilateral hearing loss after measles and rubella vaccination in an adult.
N Engl J Med. 1991

Pancreatitis caused by measles, mumps, and rubella vaccine.
Pancreas. 1991

Toxic shock syndrome: an unforeseen complication following measles vaccination.
Indian Pediatr. 1991

Perils of childhood. Immunization against measles, mumps, and rubella.
Pediatr Nurs. 1991

Risk of subacute sclerosing panencephalitis from measles vaccination.
Pediatr Infect Dis J. 1990

Central nervous system sequelae of immunization against measles, mumps,
    rubella and poliomyelitis.
Acta Paediatr Jpn. 1990

Mumps meningitis after measles, mumps, and rubella vaccination.
Lancet. 1989

Henoch-Schönlein purpura after measles immunization.
Acta Paediatr Jpn. 1989

Aseptic meningitis after vaccination against measles and mumps.
Pediatr Infect Dis J. 1989

Thrombocytopenia purpura after combined vaccine against measles, mumps,
    and rubella.
Clin Pediatr (Phila). 1987

Anaphylactic shock reaction to measles vaccine.
J R Coll Gen Pract. 1987

Diffuse retinopathy following measles, mumps, and rubella vaccination.
Pediatrics. 1985

Sensorineural hearing loss following live measles virus vaccination.
Int J Pediatr Otorhinolaryngol. 1985

Acute and delayed neurologic reaction to inoculation with attenuated live
    measles virus.
Brain Dev. 1985

Severe local reactions to live measles virus vaccine following an immunization
    program.
Am J Public Health. 1983

Convulsions after measles immunisation.
Lancet. 1983

Allergic reactions to measles (rubeola) vaccine in patients hypersensitive to egg protein.
J Pediatr. 1983

Measles virus panniculitis subsequent to vaccine administration.
J Pediatr. 1982

Immediate reactions following live attenuated measles vaccine.
Med J Aust. 1981

Thrombocytopenic purpura after measles vaccination.
N Engl J Med. 1981

Severe hypersensitivity or intolerance reactions to measles vaccine in six children. Clinical and immunological studies.
Allergy. 1980

Reye syndrome associated with vaccination with live virus vaccines. An exploration of possible etiologic relationships.
Clin Pediatr (Phila). 1979

Neurological complications following measles vaccination.
Dev Biol Stand. 1979

Sources:

Clinical and laboratory features of aseptic meningitis associated with measles-mumps-rubella vaccine Lucena R, et al., Rev Panam Salud Publica. 2002 Oct;12(4):258–61.

Haemophagocytic lymphohistiocytosis following measles vaccination. Otagiri T, et al., Eur J Pediatr. 2002 Sep;161(9):494–6. Epub 2002 Aug 8.

Gianotti-Crosti syndrome after measles, mumps and rubella vaccination. Velangi SS, et al., Br J Dermatol. 1998 Dec;139(6):1122–3.

Measles, mumps, rubella vaccine induced subacute sclerosing panencephalitis. Belgamwar RB, et al., J Indian Med Assoc. 1997 Nov;95(11):594.

Optic neuritis following measles/rubella vaccination in two 13-year-old children. Stevenson VL, et al., Br J Ophthalmol. 1996 Dec;80(12):1110–1.

Arthritis after mumps and measles vaccination. Nussinovitch M, et al., Arch Dis Child. 1995 Apr;72(4):348–9.

Hearing loss following measles vaccination. Jayarajan V, et al., J Infect. 1995 Mar;30(2):184–5.

Gait disturbances after measles, mumps, and rubella vaccine. Plesner AM, Lancet. 1995 Feb 4;345(8945):316.

Neurologic disorders after vaccination against measles and mumps. Skovránková J, et al., Cesk Pediatr. 1992 Jun;47(6):343–5.

Joint and limb symptoms in children after immunisation with measles, mumps, and rubella vaccine. Benjamin CM, et al., BMJ. 1992 Apr 25;304(6834):1075–8.

Postvaccinal parkinsonism. Alves RS, et al., Mov Disord. 1992;7(2):178–80.

Risk of seizures after measles-mumps- rubella immunization. Griffin MR, et al., Pediatrics. 1991 Nov;88(5):881–5.

Bilateral hearing loss after measles and rubella vaccination in an adult. Hulbert TV, et al., N Engl J Med. 1991 Jul 11;325(2):134.

Pancreatitis caused by measles, mumps, and rubella vaccine. Adler JB, et al., Pancreas. 1991 Jul;6(4):489–90.

Toxic shock syndrome: an unforeseen complication following measles vaccination. Phadke MA, et al., Indian Pediatr. 1991 Jun;28(6):663–5.

Perils of childhood. Immunization against measles, mumps, and rubella. Levine BE, et al., Pediatr Nurs. 1991 Mar-Apr;17(2):159–61, 215.

Risk of subacute sclerosing panencephalitis from measles vaccination. Halsey N, Pediatr Infect Dis J. 1990 Nov;9(11):857–8.

Central nervous system sequelae of immunization against measles, mumps, rubella and poliomyelitis. Ehrengut W, Acta Paediatr Jpn. 1990 Feb;32(1):8–11.

Mumps meningitis after measles, mumps, and rubella vaccination. Murray MW, et al., Lancet. 1989 Sep 16;2(8664):677.

Henoch-Schönlein purpura after measles immunization. Ozaki T, et al., Acta Paediatr Jpn. 1989 Aug;31(4):484–6.

Aseptic meningitis after vaccination against measles and mumps. Cizman M, et al., Pediatr Infect Dis J. 1989 May;8(5):302–8.

Thrombocytopenia purpura after combined vaccine against measles, mumps, and rubella. Azeemuddin S., Clin Pediatr (Phila). 1987 Jun;26(6):318.

Anaphylactic shock reaction to measles vaccine. Thurston A., J R Coll Gen Pract. 1987 Jan;37(294):41.

Diffuse retinopathy following measles, mumps, and rubella vaccination. Marshall GS, et al., Pediatrics. 1985 Dec;76(6):989–91.

Sensorineural hearing loss following live measles virus vaccination. Brodsky L, et al., Int J Pediatr Otorhinolaryngol. 1985 Nov;10(2):159–63.

Acute and delayed neurologic reaction to inoculation with attenuated live measles virus. Abe T, et al., Brain Dev. 1985;7(4):421–3.

Severe local reactions to live measles virus vaccine following an immunization program. Stetler HC, et al., Am J Public Health. 1983 Aug;73(8):899–900.

Convulsions after measles immunisation. Berlin BS., Lancet. 1983 Jun 18;1(8338):1380.

Allergic reactions to measles (rubeola) vaccine in patients hypersensitive to egg protein. Herman JJ, et al., J Pediatr. 1983 Feb;102(2):196–9.

Measles virus panniculitis subsequent to vaccine administration. Buck BE, et al., J Pediatr. 1982 Sep;101(3):366–73.

Immediate reactions following live attenuated measles vaccine. Van Asperen PP, et al., Med J Aust. 1981 Oct 3;2(7):330–1.

Thrombocytopenic purpura after measles vaccination. Kiefaber RW., N Engl J Med. 1981 Jul 23;305(4):225.

Severe hypersensitivity or intolerance reactions to measles vaccine in six children. Clinical and immunological studies. Aukrust L, et al., Allergy. 1980 Oct;35(7):581–7.

Reye syndrome associated with vaccination with live virus vaccines. An exploration of possible etiologic relationships. Morens DM, et al., Clin Pediatr (Phila). 1979 Jan;18(1):42–4.

Neurological complications following measles vaccination. Allerdist H., Dev Biol Stand. 1979;43:259–64.

Doctors Admit that Vaccines Can Cause Autoimmune Diseases

Quick Version: Auto-immune diseases are on the rise. Vaccines have been associated with some of these diseases and scientists have known this for over twenty-five years.

It's clear that some children are getting injured or killed by vaccines. It's also clear that many injuries are going unreported.

There is also another phenomenon that some doctors have linked to vaccines—the rise of autoimmune diseases. More people are getting these diseases than ever before.

The NIH admits autoimmune diseases are on the rise and there is probably an environmental trigger.
The National Institute of Health (NIH) says:

> More than 80 diseases occur as a result of the immune system attacking the body's own organs, tissues, and cells. Some of the more common autoimmune diseases include type 1 diabetes, rheumatoid arthritis, systemic lupus erythematosus, and inflammatory bowel disease. Collectively, autoimmune diseases are among the most prevalent diseases in the U.S., affecting more than 23.5 million Americans.

According to Frederick W. Miller, MD, PhD, director of the Environmental Autoimmunity Group at the National Institute of Environmental Health Sciences at the NIH, autoimmune diseases are rising largely because of environmental reasons. Miller states, "[W]e've got 80,000 chemicals approved for use in commerce, but we know very little about their immune effects. Our lifestyles are also different than they were a few decades ago, and we're eating more processed food."

What "environmental reasons" are causing these autoimmune disease rates to rise so quickly? Is it processed foods and the thousands of chemicals in our air, water, and soil? Can vaccines also be playing a role?

The rates of some autoimmune diseases are sky-rocketing.

Let's look at the rates of some auto-immune diseases from various sources. The data are often difficult to obtain. When we reached out to the CDC we were told there is no comprehensive database. Strange.

Type 1 Diabetes
There was a 21 percent increase in people diagnosed with Type 1 diabetes between 2001 and 2009 under the age of twenty.

Systematic Lupus Erythematosus
Rate has nearly doubled in one generation.
1988–1994: 53.6/100,000 people per year
2003–2008: 102.9/100,000 people per year

Multiple Sclerosis
Rate has skyrocketed.
1975: 123,000 cases
2017: over 850,000 cases

Crohn's Disease
Almost a 70 percent increase in one generation.
1991: ~140/100,000 people per year
2008–2009: 235.6/100,000 people per year

Guillain-Barré Syndrome
Almost doubled in thirty-four years.
1973: 1.6/100,000 people per year
2007: 2.9/100,000 people per year

Sarcoidosis
Almost doubled in forty-five years.
1967–1987: 4.8/100,000 people per year
2010: 8.4/100,000 people per year
2012: 8.8/100,000 people per year

According to epidemiologists (scientists who study disease), the long-term effects of individual vaccines and of the vaccination program itself remain unknown. Yet, vaccines are being promoted as "safe" to every child in the United States.

An Italian study says vaccines may "potentially even trigger a full-blown autoimmune disease."
A study from Italy looked at autoimmune diseases reported after vaccination. The table below summarizes some of the studies they reviewed that demonstrate a potential relationship between vaccination and some autoimmune diseases.

Autoimmune Disease	Type of vaccine
Systemic lupus erythematosus	HBV, tetanus, anthrax
Rheumatoid arthritis	HBV, tetanus, typhoid, parathypoid, MMR
Multiple sclerosis	HBV
Reactive arthritis	BCG, typhoid, DPT, MMR, HBV influenza
Guillain-Barré syndrome	Influenza, polio, tetanus
Diabetes mellitus-type 1	HIB
Idiopathic thrombocytopenia	MMR, HBV
Hashimoto thyroiditis	HBV

Below are some of the specific findings from this study with the translation in non-medical language:

- "In the 1994, Stratton and coworkers published the first report on a causal relationship between several vaccines (e.g., diphtheria, tetanus toxoids, oral polio vaccines) and autoimmune disorders (e.g., Guillain–Barre syndrome, type 1 diabetes, and multiple sclerosis)." *Translation: Scientists found that vaccination caused some autoimmune diseases in 1994!*
- "These autoimmune disorders (rheumatic, endocrinological, and gastrointestinal diseases) are increased significantly over the last 30

years and affect more than 5% of the individuals worldwide at the
age of vaccination programs, which is quite different compared to
the spontaneous autoimmune disease incidence."
*Translation: Certain autoimmune diseases have increased a lot in thirty
years, especially among people who were vaccinated.*
 • Their conclusions and expert recommendations included that
   "Vaccination might display autoimmune side effects and potentially
   even trigger a full-blown autoimmune disease."

The US government knew that a vaccine-autoimmune link existed in 1991.

As outrageous as this sounds, many years ago, in 1991, the US government
knew of the relationship between autoimmune diseases and vaccines.

In 1991, the government's own Institute of Medicine released a report
entitled "Adverse Effects of Pertussis and Rubella Vaccines." In that report,
the committee stated that the evidence indicates a causal relation between
the rubella vaccine and acute arthritis in adult women.

Two years later, in September 1993, the Institute of Medicine released
another report entitled "Adverse Events Associated With Childhood
Vaccines: Evidence Bearing on Causality." These scientists wrote, "The
committee found that the evidence favored acceptance of a causal relation
between:

 • Diphtheria and tetanus toxoids (vaccine) and Guillain-Barré
   syndrome and brachial neuritis.
   AND
 • Oral polio vaccine and Guillain-Barré syndrome,
   Guillain-Barré syndrome is an autoimmune disease in which
   the body's immune system attacks the body's nerves. Weakness
   and tingling in the extremities are usually the first symptoms.
   These sensations quickly spread, eventually paralyzing the person.
   Remember, the rate of Guillain-Barré syndrome has almost doubled
   in thirty-four years. In 1993 the government said that some vaccines
   can cause this autoimmune disease.

Today, the government is telling us that vaccines do not cause autoimmune
diseases, yet their own reports said the opposite over twenty-five years ago;
some vaccines could cause some autoimmune diseases.

This does not mean that all auto immune diseases are caused by vaccines. Dr. Miller of the NIH is probably right when he says that the eighty thousand chemicals approved for use in commerce, and our lifestyles may be playing a role in the rise of autoimmune diseases. However, when scientists discuss environmental reasons for the increase of a disease, why not consider vaccines? Getting exposed to viruses, bacteria, chemicals and heavy metals in a syringe is yet another form of environmental exposure.

Sources:

"Research Published in Other Medical Journals have Linked Diabetes to Various Vaccines including: Hemophilus Vaccine, Pertussis, MMR, and BCG Vaccine." *Endocrinology and Metabolism* 16, no. 4 (April/May 2003), 495–508.

"Autoimmune Diseases and Your Environment." National Institute of Environmental Health Sciences (NIEHS). Accessed February 11, 2021. https://www.niehs.nih.gov/health /materials/autoimmune_diseases_508.pdf.

"Autoimmune Diseases." National Institute of Allergy and Infectious Disease. Accessed February 11, 2021. https://www.niaid.nih.gov/diseases-conditions/autoimmune-diseases.

"The Autoimmune Epidemic Excerpt." *Donna Jackson Nakazawa.* Accessed February 11, 2021. https://donnajacksonnakazawa.com/the-autoimmune-epidemic-excerpt/.

Baughman, Robert P., and Shelli Field. "Sarcoidosis in America. Analysis Based on Health Care Use." *Annals of the American Thoracic Society* 13, no. 8 (2016), 1244–1252. doi:10.1513/annalsats.201511-7600c.

Baum, Herbert M., and Beth B. Rothschild. "The incidence and prevalence of reported multiple sclerosis." *Annals of Neurology* 10, no. 5 (November 1981), 420–428. doi:10 .1002/ana.410100504.

Beghi, Ettore. "Guillain-Barré Syndrome: Clinicoepidemiologic Features and Effect of Influenza Vaccine." *Archives of Neurology* 42, no. 11 (1985), 1053–1057. doi:10.1001/archneur .1985.04060100035016.

Classen, J.B., and D.C. Classen. "Clustering of Cases of Type 1 Diabetes Mellitus Occurring 2–4 Years After Vaccination is Consistent with Clustering After Infections and Progression to Type I Diabetes Mellitus in Autoantibody Positive Individuals." *Journal of Pediatric Endocrinology and Metabolism* 16, no. 4 (2003), 1218–1226.

Dilokthornsakul, Piyameth, Robert J. Valuck, Kavita V. Nair, John R. Corboy, Richard R. Allen, and Jonathan D. Campbell. "Multiple sclerosis prevalence in the United States commercially insured population." *Neurology* 86, no. 11 (March 2016), 1014–1021. doi:10.1212/wnl.0000000000002469.

Geiss, Linda S., Jing Wang, Yiling J. Cheng, Theodore J. Thompson, Lawrence Barker, and Yanfeng Li. "Prevalence and Incidence Trends for Diagnosed Diabetes Among Adults Aged 20 to 79 Years, United States, 1980–2012." *JAMA* 312, no. 12 (2014), 1218–1226.

Helmick, Charles G., David T. Felson, and Reva C. Lawrence. "Estimates of the prevalence of arthritis and other rheumatic conditions in the United States: Part I." *Arthritis & Rheumatism* 58, no. 1 (2007), 15–25.

Howson, Christopher P. "Adverse Events Following Pertussis and Rubella Vaccines." *JAMA* 267, no. 3 (January 1992), 392–396. https://www.ncbi.nlm.nih.gov/pubmed/1727962

Kappelman, Michael D., Kristen R. Moore, Jeffery K. Allen, and Suzanne F. Cook. "Recent Trends in the Prevalence of Crohn's Disease and Ulcerative Colitis in a Commercially Insured US Population." *Digestive Diseases and Sciences* 58, no. 2 (2012), 519–525. doi:10.1007/s10620-012-2371-5.

Mayer-Davis, Elizabeth J. "Incidence Trends of Type 1 and Type 2 Diabetes Among Youths, 2002–2012 | NEJM." *New England Journal of Medicine*. Last modified April 12, 2017. https://www.nejm.org/doi/full/10.1056/NEJMoa1610187.

Nelson, Laura, Robert Gormley, Mark S. Riddle, David R. Tribble, and Chad K. Porter. "The epidemiology of Guillain-Barré Syndrome in U.S. military personnel: a case-control study." *BMC Research Notes* 2, no. 1 (August 2009), 171.

"Rates of New Diagnosed Cases of Type 1 and Type 2 Diabetes on the Rise Among Children, Teens." National Institutes of Health (NIH). Last modified April 13, 2017. https://www.nih.gov/news-events/news-releases/rates-new-diagnosed-cases-type-1-type-2-diabetes-rise-among-children-teens.

Schmidt, Charles W. "Questions Persist: Environmental Factors in Autoimmune Disease." *Environmental Health Perspectives* 119, no. 6 (June 2011), A248-A253. doi:10.1289/ehp.119-a248. https://www.ncbi.nlm.nih.gov/pmc/articles/PMC3114837/.

Shivashankar, Raina, William J. Tremaine, W. S. Harmsen, and Edward V. Loftus. "Incidence and Prevalence of Crohn's Disease and Ulcerative Colitis in Olmsted County, Minnesota From 1970 Through 2010." *Clinical Gastroenterology and Hepatology* 15, no. 6 (2017), 857–863.

Stojan, George, and Michelle Petri. "Epidemiology of systemic lupus erythematosus: an update." *Current Opinion in Rheumatology* 30, no. 2 (2018), 144–150. doi:10.1097/bor.0000000000000480.

Stratton, Kathleen R. "Adverse Events Associated With Childhood Vaccines Other Than Pertussis and Rubella—A Summary of a Report from the Institute of Medicine." *JAMA* 271, no. 20 (May 1994), 1602. https://www.ncbi.nlm.nih.gov/pubmed/8182813.

Stratton, Kathleen R. *Adverse Events Associated with Childhood Vaccines: Evidence Bearing on Causality*. Washington, DC: National Academies Press, 1993. Presentation.

Thomeer, M., M. Demedts, and W. Wuyts. "Epidemiology of sarcoidosis." *European Respiratory Journal*, 2005, 13–22.

"Type 1 Diabetes Statistics." Beyond Type 1. Last modified August 20, 2020. https://beyondtype1.org/type-1-diabetes-statistics/.

Vadalà, Maria, Dimitri Poddighe, Carmen Laurino, and Beniamino Palmieri. "Vaccination and autoimmune diseases: is prevention of adverse health effects on the horizon?" *EPMA Journal* 8, no. 3 (2017), 295–311.

Vehik, K., and R. F. Hamman. "Increasing Incidence of Type 1 Diabetes in 0- to 17-Year-Old Colorado Youth." *Diabetes Care* 30, no. 3 (2007), 503–509.

Wallin, Mitchell T., William J. Culpepper, Jonathan D. Campbell, Lorene M. Nelson, Annette Langer-Gould, Ruth A. Marrie, Gary R. Cutter, et al. "The prevalence of MS in the United States." *Neurology* 92, no. 10 (2019), e1029-e1040. doi:10.1212/wnl .0000000000007035.

Safety Is in the Eye of the Beholder

> *Quick Version:* People in the vaccine business know that some kids may be seriously injured or killed from vaccines, but, in their judgment, it's worth the risk. After all, it's not their kids.

If the government tells us a vaccine is safe, what does that mean?

1. It is safe for every child.
   Or
2. That some kids may be seriously injured or killed, but, in their judgment, it's worth the risk.

In fact, the answer is number two.

But who decides that it's worth the risk? Why should politicians and bureaucrats decide when it's your child, not theirs?

The definition of something that is "safe" is "free from harm or risk," according to the *Merriam-Webster Dictionary*.

However, when scientists talk about safety, they don't mean that something is absolutely (100 percent) safe. They mean something else. It's something much more subjective. It's a weighing of pros and cons. It's a balancing act.

RotaShield was pulled off the market because it killed.

Here's an example of a vaccine that the authorities determined was not safe.

RotaShield was a vaccine designed to protect children against a virus that causes diarrhea. It was approved by the FDA on August 31, 1998. Distribution began on October 1, 1998.

By May 1999, life-threatening intussusception injuries were reported.

Intussusception is where one part of intestine "telescopes" inside of another, causing a blockage. It usually occurs at the connection of the small and large intestines. It is dangerous and can kill the person.

The number of these injuries continued to increase and there was at least one death (a five-month girl).

Wyeth-Lederle, the manufacturer of the vaccine, suspended further distribution of this vaccine on July 16, 1999, and withdrew it from the market on October 15, 1999.

According to a Congressional investigation, the vaccine's clinical trials *prior* to approval demonstrated an intussusception rate of five in 10,054 children.

Based on injecting 3,800,000 children, the number of intussusception injuries would be about 1,865 per year. Such a large number of injuries and potential deaths was much bigger than whatever benefit the vaccine might provide in preventing diarrhea. Nonetheless, the bureaucrats at the FDA and CDC decided that the vaccine was safe. (As we have seen in earlier chapters, many of these decision makers had a personal financial interest in getting the vaccine to market.)

The studies (before the vaccine was released) also demonstrated concerns about children "failing to thrive," developing high fevers which can lead to brain injury, and experiencing slowing of growth. But, again, the health authorities decided this vaccine was safe.

Although the government was aware of this information, they approved the vaccine. But, after the intussusception injuries went public, their approval was withdrawn.

Vaccine doctor explains that "absolute safety" is not "reasonable."

But revoking this vaccine had its critics. One was Paul Offit. He is a pediatrician who has made millions of dollars from vaccines. He co-invented another rotavirus vaccine for which he receives "an unspecified sum of money for his interest." He has also been a member of the CDC advisory committee that recommends vaccines for the entire country. He recommended the approval of RotaShield, the vaccine that killed and was pulled off the market.

This is what he said about pulling RotaShield from the CDC's list of recommended vaccines:

> We said this is unsafe for American children, period, without ever defining safety. What we meant by doing it the way we did was absolute safety, which isn't a reasonable definition. It's a lawyer's definition. It's not a doctor or scientist's definition" (Offit 2016).

In English: Offit is admitting that "absolute safety" is the wrong standard for a vaccine. ("It's a lawyer's definition. It's not a doctor or scientist's definition.") Some other standard is needed.

According to Jason L. Schwartz, assistant professor in the Department of Health Policy and Management at the Yale School of Public Health, "subjective assessments" have to be made when deciding whether to approve a vaccine.

What does all this mean?

It means that doctors and scientists know that vaccines are not safe for every child. It also means that deciding to mandate a vaccine like measles or any other childhood vaccine is subjective.

So, here's the problem in a nutshell:

A subjective decision will be made and people and companies who profit from vaccines will always exaggerate the benefits and minimize the risks of vaccination.

If vaccines are not absolutely safe for every child, then why should these people, the ones with money interests, be deciding to mandate vaccines for your child? Do they know your child? Do they love your child?

If vaccines are not safe, then why shouldn't parents decide?

Doesn't it make sense that the person making medical decisions for a child be a person who loves that child? Or do you think it is better that a bureaucrat or politician or someone making money from vaccines makes these decisions for you?

With mandatory vaccination, parents don't count. The person making the subjective decision that the vaccine is "safe enough" and your child must be vaccinated is not you. These bureaucrats apply their own subjective standards and many of these people are enriching themselves from promoting vaccines.

This is called conflict of interest.

How do we combat conflicts of interest? The easiest way is to allow parents to decide whether vaccination is appropriate for a given child. Parents love their children so let them protect their children. Let parents weigh the pros and cons of vaccines with the help of their doctor and make an informed decision.

Sources:

Horwin, Michael. "Ensuring Safe, Effective and Necessary Vaccines for Children." *CWSL Scholarly Commons | California Western School of Law Research.* Last modified 2001. https://scholarlycommons.law.cwsl.edu/cgi/viewcontent.cgi?article=1203&context =cwlr.

Schwartz, Jason. "The First Rotavirus Vaccine and the Politics of Acceptable Risk." PubMed Central (PMC). Accessed February 11, 2021. https://www.ncbi.nlm.nih.gov /pmc/articles/PMC3460207/.

Anna Kirkland. "Vaccine Court: The Law and Politics of Injury," New York University Press 2016.

PART 4

The Shocking Ingredients in the Measles Vaccine

Do you know what exactly is in the syringe being injected into your child? Do you know where the vaccine comes from or how it is made? We are constantly being told to focus on the potential benefit of vaccines. This is understandable; after all, we do this with everything we consume—focus on the intended use or benefit. But most products are not injected into our children's bodies. Since vaccines are, shouldn't we know what is in those syringes?

Secrets:
20. The Disturbing Way the Measles Vaccine Was Created
21. The Measles Vaccine Is Grown on Tissue from Aborted Human Fetuses
22. The Measles Vaccine Is Made from Chicken Embryo Cell Culture that May Contain Bird Virus Material
23. The Measles Vaccine Contains Over a Dozen Other Ingredients and Chemicals Including One that Is Genetically Engineered

The Disturbing Way the Measles Vaccine Was Created

> ***Quick Version:*** The measles vaccine used today was passed through a total of ninety-eight individual chicken and human cell cultures. The testing on the purity of this product depends on decades-old systems.

How viral vaccines are created

The measles vaccine uses a live measles virus. However, if you administer live measles virus, children will get measles.

Therefore, scientists wanted a way to create a weakened strain—a virus that could create an immune response, but not cause measles. This process of taking a normal virus (what scientists call "wild type") and creating something less dangerous for a vaccine is called "attenuation."

To attenuate or weaken a virus you have to pass it through other types of cells. The idea is that the virus will change slightly with each passing (called "passages") and will get less virulent or dangerous.

Vaccines created this way are called LAV vaccines. LAV stands for "live attenuated viral" vaccines.

- It's live because the virus is actually living.
- It's attenuated because it weakened through passaging.

- It's viral because it's not a bacteria. It's a virus.

LAV vaccines include: measles, mumps, rubella, vaccinia (smallpox), varicella (chickenpox), zoster (which contains the same virus as varicella vaccine but in much higher amount), yellow fever, rotavirus, and influenza (intranasal).

To create the measles LAV, scientists in the 1950s and 1960s experimented with lots of different tissues to passage the virus. Remember, passaging means putting the virus into certain cells and then harvesting the virus from the cells after the virus has replicated. It's like planting a seed in the ground, but instead of a seed it's a virus and instead of the ground it's a cell culture.

Here's the problem. When you use this method, you can end up with ("harvest") other viruses besides measles. Hundreds and perhaps thousands of viruses infect various human and animal tissues. When you use those tissues as your way of weakening and growing the virus you want, you may end up with viruses you don't want.

Oral polio vaccine and cancer-causing monkey virus

This is exactly what happened in the case of the oral polio vaccine when it became contaminated with a cancer-causing monkey virus called SV40 (Simian Virus 40). Scientists had put the polio vaccine into monkey kidney cells to weaken it (attenuation) and when they harvested the polio virus from the monkey kidneys, they also harvested dangerous monkey viruses.

They also used monkey kidney cells to grow the vaccine and ran into the same problem. Hundreds of millions of people throughout the world, including the United States, were vaccinated with polio vaccines contaminated with the cancer-causing monkey virus, SV40. Today, SV40 has been found in the cancers of many children and adults who got the oral polio vaccine.

In fact SV40 was the fortieth monkey virus found in polio vaccine. There were hundreds of other money viruses that were found in the vaccine. It's not surprising. Kidneys are the organs that remove toxic substances and help keep the blood clean. If you grow a vaccine in monkey kidneys you will likely harvest other monkey viruses. Grow a vaccine in human kidneys and you will likely harvest other human viruses.

Creation of the measles vaccine

Returning to the measles, here is how this vaccine was created, the so-called Edmonston B strain.

Back in 1954, scientists John Enders and Thomas Peebles isolated the measles virus. Here's how they describe what they did:

> [In] 1954 [we] undertook experiments in which cultures of human postnatal tissues in roller tube cultures were exposed to whole blood and throat washings obtained from a patient with measles during the first 24 hours of the exanthem.

Translation: They took cells from a woman after her child was born. That's what postnatal means. It's most likely amnion cells (the innermost membrane that encloses the embryo of a human baby) because these cells are used later. But they don't actually say that. Then they infected these cells with blood and throat cultures from another person who had measles. This blood and culture was collected within twenty-four hours after the patient had a rash. (Exanthem means rash).

The person whose blood and throat cultures they used was an eleven-year old boy named David Edmonston. Why is that a fair assumption? Because later this measles strain would be called "Edmonston."

Passaging the measles virus through human and chicken cells

But the scientists aren't done yet. Now that they have captured the measles virus, they have to weaken it by passing it through other cells. This is what they did:

- Twenty-four passages through human renal (kidney) cells
- Twenty-eight passages through primary cultures of human amnion cells (the innermost membrane that encloses the embryo of a human baby)
- Six passages through chicken embryo cells

After these fifty-eight passages through other cells of humans and chickens they now had the Edmonston B strain. This was the source of the measles virus for the measles vaccine and would be used until 1975.

After 1975, Edmonston B underwent an additional forty passages in chicken embryo cells to create the Moraten strain, the one in use today. Moraten simply meant "more attenuated Enders."

So the measles vaccine used today was passed through a total of ninety-eight chicken and human cell cultures.

These are the master seed strains that were used to make all the measles vaccines for the entire country.

Did they pick up any other viruses along the way? Where are the hi-tech tests to show that no advantageous agents (extra viruses) are present?

Remember that the technology to make the viral part of this vaccine is now over sixty years old. In those days, there were no high-tech tests like PCR DNA to find contaminants or extraneous viruses.

Unfortunately, it doesn't seem to matter to the CDC or FDA or Merck. We are still using a vaccine which is the product of six-decades-old research with little, if any, updated testing for contaminants or extraneous viruses or viral material.

Sources:

Enders, John F. "Development of Attenuated Measles Virus Vaccines." *American Journal of Diseases of Children* 103, no. 3 (March 1962), 335–340.

Enders, John F., Thomas C. Peebles, Kevin McCarthy, Milan Milovanović, Anna Mitus, and Ann Holloway. "Measles Virus: A Summary of Experiments Concerned with Isolation, Properties, and Behavior." *American Journal of Public Health and the Nations Health* 47, no. 3 (March 1957), 275–282. doi:10.2105/ajph.47.3.275.

Karelitz, Samuel. "Measles Vaccine." *JAMA* 177, no. 8 (August 1961), 537–541.

"Live Attenuated Measles Vaccine." *EPI Newsletter* 2, no. 1 (February 2008), 6.

The Measles Vaccine Is Grown on Tissue from Aborted Human Fetuses

Quick Version: The MMR and MMRV vaccines were both grown on tissue from aborted human fetuses and residual components of these cells, including DNA and protein, are in the vaccine.

After the measles vaccine was isolated and attenuated as described above, the scientists had a very small amount of the weakened measles virus. How could they embark on a nationwide vaccine campaign and inoculate millions of children?

Growing the measles vaccine in tissue from aborted fetuses

They needed a way to grow more. They needed a substrate.

What's a substrate?

It is some type of growth medium for a virus. An example would be a cell culture, cells from another living organism.

Today, measles is part of the measles-mumps-rubella vaccine (MMR) or the measles-mumps-rubella-varicella (chickenpox) vaccine MMRV (called ProQuad). Since all four vaccines are live viruses, they all faced the same challenge—how to grow more?

These are the substrates chosen:

- For measles and mumps they used chick embryo cell cultures.
- For rubella they used WI-38 human diploid lung fibroblasts from a three-month gestation aborted female (human) fetus.
- For the varicella (chickenpox vaccine) they used a MRC-5 diploid human cell culture line from fibroblasts derived from lung tissue of a fourteen-week-old aborted Caucasian male (human) fetus.

Therefore, if your child gets ProQuad (MMRV) they are getting injected with a vaccine grown on the tissue of two different aborted fetuses (male and female).

If your child only gets MMR, they are getting injected with a vaccine grown on the tissue of one aborted fetus (female).

Some people claim that no remnants of the aborted fetal tissue get into the actual vaccine.

That's false.

Let's take this step by step.

First, let's look at what Merck says in their vaccine inserts that describe their products. It's a little hard to read so if you would like, just pick out the words we bolded. These are the cell substrates, how the viruses in the vaccine were grown:

M-M-R II (Measles, Mumps, and Rubella Virus Vaccine Live) is a live virus vaccine for vaccination against measles (rubeola), mumps, and rubella (German measles). M-M-R II is a sterile lyophilized preparation of:
- (1) ATTENUVAX (Measles Virus Vaccine Live), a more attenuated line of measles virus, derived from Enders' attenuated Edmonston strain and propagated in **chick embryo cell culture**;
- (2) MUMPSVAX (Mumps Virus Vaccine Live), the Jeryl Lynn (B level) strain of mumps virus propagated in **chick embryo cell culture**; and
- (3) MERUVAX II (Rubella Virus Vaccine Live), the Wistar RA 27/3 strain of live attenuated rubella virus propagated in **WI-38 human diploid lung fibroblasts**.

ProQuad (Measles, Mumps, Rubella and Varicella Virus Vaccine Live) is a combined, attenuated, live virus vaccine containing measles, mumps, rubella, and varicella viruses.

ProQuad is a sterile lyophilized preparation of all of the above for the measles, mumps and rubella portion plus "Varicella Virus Vaccine Live (Oka/

Merck), the Oka/Merck strain of varicella-zoster virus propagated in **MRC-5 cells**.

Remember that WI-38 human diploid lung fibroblasts come from a three-month-gestation aborted female (human) fetus. This is what Wikipedia says:

> "WI-38 is a diploid human cell line composed of fibroblasts derived from lung tissue of a 3-month-gestation female fetus. The fetus came from the elective abortion of a Swedish woman in 1962, and was used without her knowledge or permission."

And this is what Wikipedia says about MRC-5 cells:

> "MRC-5 (Medical Research Council cell strain 5) is a diploid cell culture line composed of fibroblasts, originally developed from the lung tissue of a 14-week-old aborted Caucasian male fetus."

Yes, remnants of these aborted fetus cell lines are in the vaccine.

Now, the big question, does any of that aborted human tissue get into the vaccine?

Vaccine proponents point out that just because a vaccine is grown on tissue from aborted human fetuses or chick embryos, it doesn't mean that parts of this tissue make their way into the final vaccine product. Let's take a look.

Measles vaccine bulk is an unpurified product.

The European Medicines Agency (EMA) is the agency of the European Union responsible for the scientific evaluation, supervision, and safety monitoring of medicines in the European Union.

In 2006, the agency published their findings about the measles vaccine made by Merck (MSD in Europe):

> Measles vaccine bulk is an unpurified product . . . Degradation products are neither identified nor quantified. Process-related impurities arising from the measles vaccine bulk manufacturing processes . . . may include proteins derived from the host organism . . . antibiotics (e.g., neomycin), serum, or other media components.

In plain English, this means that the vaccine is not purified. It contains degradation products such as unwanted substances that can develop during the manufacturing of the vaccine. What are these products?

Merck says that residual components of human aborted fetal cells including DNA and protein are in the vaccine.

Merck admits that parts of cells from an aborted human fetus are in the vaccine.

In their insert for measles-mumps-rubella-chickenpox vaccine (ProQuad), the manufacturer states:

> Each 0.5-mL dose of the vaccine nominally contains 20 mg of sucrose, 11 mg of hydrolyzed gelatin, 2.5 mg of urea; 2.3 mg of sodium chloride, 16 mg of sorbitol, 0.38 mg of monosodium L-glutamate, 1.4 mg of sodium phosphate, 0.25 mg of recombinant human albumin, 0.13 mg of sodium bicarbonate, 94 mcg of potassium phosphate, 58 mcg of potassium chloride; **residual components of MRC-5 cells including DNA and protein**; 5 mcg of neomycin, bovine serum albumin (0.5 mcg), and other buffer and media ingredients" (emphasis added).

Remember that MRC-5 cells are derived from lung tissue of a fourteen-week-old aborted Caucasian (human) male fetus. So "residual components . . . including DNA and protein" from this fetus are in the final vaccine. This is what the manufacturer tells us.

These dots are easy to connect.

Sources:
"Highlights of Prescribing Information M-M-R II." *Merck*. Accessed February 11, 2021. https://www.merck.com/product/usa/pi_circulars/m/mmr_ii/mmr_ii_pi.pdf.
"Highlights of Prescribing Information—ProQuad." U.S. Food and Drug Administration. Accessed February 11, 2021. https://www.fda.gov/media/75195/download.
"M-M-R® II(MEASLES, MUMPS, and RUBELLA VIRUS VACCINE LIVE)." World Health Organization. Accessed February 11, 2021. https://www.who.int/immunization _standards/vaccine_quality/PQ_168_MMR_MSD_PI_July2008.pdf.
"MRC-5." Wikipedia. Last modified May 19, 2016. https://en.wikipedia.org/wiki/MRC-5.
"ProQuad Product Monograph." hres.ca. Accessed February 11, 2021. https://pdf.hres.ca /dpd_pm/00037770.PDF.
"WI-38." Wikipedia. Last modified December 17, 2012. https://en.wikipedia.org/wiki /WI-38.

The Measles Vaccine Is Made from Chicken Embryo Cell Culture that May Contain Bird Virus Material

Quick Version: Both the measles and mumps part of the MMR vaccine are grown in chicken embryo cell culture. Studies have found bird viruses in the culture. Your chance of getting a bird virus in the vaccine that can infect you and spread may depend on chance.

Both the measles and mumps part of the MMRII vaccine are grown in chicken embryo cell culture. Several studies have reported finding avian viruses in the culture. These viruses include endogenous avian leukosis viruses and endogenous avian viruses.

What are these?

These are viruses that can cause tumors in chickens and other birds.

Children who get the measles vaccine are exposed to chicken retroviral particles.

According to one study, "Analysis of MMR vaccines from different manufacturers suggests that vaccine recipients may be *universally exposed* to endogenous chicken retroviral particles."

In other words, people are getting pieces of these bird viruses injected into them from the vaccine.

Perhaps the most worrisome part of that statement is the term "universally exposed." That is understood to mean everybody.

A few limited studies have been performed to find out if these bird viruses are spreading in people who get the vaccine. So far, thankfully, they have not been found to be spreading.

However, the last study only looked at 206 vaccine recipients out of the millions who get the vaccine. The researchers said: "The proportion of defective to infectious Avian leukosis virus in different vaccine lots depends on the . . . chick embryo fibroblast substrate preparation used for each vaccine lot."

In other words, your risk of getting an infectious bird virus—one that has a virus that can reproduce and spread—depends on which lot of vaccines you get and how it was prepared. It's luck of the draw.

Sources:

Hussain, Althaf. "Lack of Evidence of Endogenous Avian Leukosis Virus and Endogenous Avian Retrovirus Transmission to Measles Mumps Rubella Vaccine Recipients." *Emerging Infectious Diseases* 7, no. 1 (January/February 2001), 66–72.

Johnson, Jeffrey A., and Walid Heneine. "Characterization of Endogenous Avian Leukosis Viruses in Chicken Embryonic Fibroblast Substrates Used in Production of Measles and Mumps Vaccines." *Journal of Virology* 75, no. 8 (April 2001), 3605–3612.

The Measles Vaccine Contains Over a Dozen Other Ingredients and Chemicals Including One that Is Genetically Engineered

Quick Version: The measles vaccine contains a brew of many different substances and chemicals including one that is genetically engineered. Where are the long-term safety tests?

In addition to bird viral particles in the vaccine, egg proteins, and DNA from aborted human fetuses, here are the other vaccine ingredients listed by the manufacturer:

Each 0.5-mL dose of the MMRII vaccine contains:

- Sodium Phosphate, Monobasic 3.1 mg.—This is a common industrial chemical. Phosphates are often used in foods and in water treatment. It is added in animal feed, toothpaste, and evaporated milk. It is used as a thickening agent and emulsifier.
- Sodium Phosphate, Dibasic 2.2 mg.—This is similar to Monobasic above.
- Sodium Bicarbonate 0.5 mg.—Sodium bicarbonate can be an effective way of controlling fungal growth and in the United

States is registered by the Environmental Protection Agency as a
biopesticide. Sodium bicarbonate mixed with water can be used as
an antacid to treat acid indigestion and heartburn.

- Medium 199 3.3 mg.—Medium 199 was the first nutritionally
  defined medium developed by Morgan, Morton, and Parker in 1950.
  This complex medium was formulated specifically for nutritional
  studies on primary chick embryo fibroblasts in the absence of any
  additives.
- Minimum Essential Medium, Eagle 0.1 mg.—A synthetic cell
  culture medium developed by Harry Eagle first published in 1959
  that can be used to maintain cells in tissue culture.
- Neomycin 25.0 µg.—Neomycin is an antibiotic found in many
  topical medications such as creams, ointments, and eye drops.
- Phenol Red 3.4 µg.—Phenol red is a pH indicator frequently used
  in cell biology laboratories and swimming pool test kits.
- Sorbitol 14.5 mg.—Is a sugar alcohol with a sweet taste which the
  human body metabolizes slowly.
- Gelatin (Porcine) Hydrolyzed 14.5 mg.—gelatin from a pig.
- Sucrose 1.9 mg.—sugar.
- Monosodium L-Glutamate 20.0 µg.—a flavor enhancer and
  stabilizer linked to headaches in some people.
- Recombinant human albumin (rHA)—rHA is genetically
  engineered and made by inserting the human gene for albumin in
  yeast cells or rice. Human albumin is the most abundant protein in
  plasma (the clear or yellowish part of blood). rHA protects vaccines
  from aggregation, oxidation, and improves solubility.

Since rHA is genetically engineered, has it been thoroughly tested for many
years to assess its safety? No. Has it been tested when injected into children
in combination with viruses and other chemicals that are in vaccines? No.
Have any of these other substances been independently tested in double
blind placebo controlled studies? No.

If you were to swallow these substances it would not be healthy, but at
least your stomach acid could break down and digest some of these chemicals. Unfortunately, by injecting them directly into the body of your child
and bypassing their digestive system, there is little defense that can be
mounted.

Sources:

"Eagle's Minimal Essential Medium." Wikipedia. Last modified October 27, 2004. https://en.wikipedia.org/wiki/Eagle%27s_minimal_essential_medium.

"Media 199." Biological Industries USA—Culture of Excellence. Accessed February 11, 2021. https://www.bioind.com/products/cell-culture/classical-media/media-199/.

"Medium 199With Earle's salts and 25mM HEPES Bithout L-Glutamine and Sodium Bicarbonate." HiMedia Leading BioSciences Company. Accessed February 11, 2021. https://himedialabs.com/TD/AT094A.pdf.

"M-M-R® II(MEASLES, MUMPS, and RUBELLA VIRUS VACCINE LIVE)." World Health Organization. Accessed February 11, 2021. https://www.who.int/immunization_standards/vaccine_quality/PQ_168_MMR_MSD_PI_July2008.pdf.

"Monosodium Phosphate." Wikipedia. Last modified May 17, 2007. https://en.wikipedia.org/wiki/Monosodium_phosphate.

"Neomycin." Wikipedia. Last modified June 28, 2003. https://en.wikipedia.org/wiki/Neomycin.

"Phenol Red." Wikipedia. Last modified January 25, 2005. https://en.wikipedia.org/wiki/Phenol_red.

"Sodium Bicarbonate." Wikipedia. Last modified December 8, 2002. https://en.wikipedia.org/wiki/Sodium_bicarbonate#Medical_uses_and_health.

"Sorbitol." Wikipedia. Last modified June 12, 2002. https://en.wikipedia.org/wiki/Sorbitol.

PART 5

The Autism-Vaccine Secrets You Have Never Been Told

Secrets:

24. Government Scientists and Doctors Admit the Vaccine-Autism Connection Is Real and Deserves Further Study (But the Government Doesn't Want These Studies)
25. US Government Data Describes Children Becoming Autistic After Getting Vaccinated
26. US Government Pays Families After Child Gets Encephalitis from Vaccines (But the Description of Encephalitis Sounds Like Autism)
27. Studies Have Found a Biological MMR Vaccine-Autism Connection
28. The Vaccine-Autism Connection Is Being Ignored and Denied Because It Would Cost the US Government Trillions of Dollars

Government Scientists and Doctors Admit the Vaccine-Autism Connection Is Real and Deserves Further Study (But the Government Doesn't Want These Studies)

Quick Version: The former head of the National Institute of Health says that we have to take another look at vaccines and autism. The government has paid about twenty million dollars to a child who got autism after vaccines. The CDC's director of Immunization Safety says it's hard to predict which children might be susceptible to autism from vaccines. A CDC whistleblower said that the CDC did not report significant findings about autism and vaccines. A study found that four times more vaccinated children are autistic.

We have all heard about autism. What is it?

The facts about autism

The following is from the National Institute of Health:

Autism spectrum disorder (ASD) refers to a group of complex neurodevel-
opment disorders characterized by repetitive and characteristic patterns of
behavior and difficulties with social communication and interaction. The
symptoms are present from early childhood and affect daily functioning.

In 2018, the CDC determined that approximately one in fifty-nine children
are diagnosed with an autism spectrum disorder.

- One in thirty-seven boys.
- One in 151 girls.
- Boys are four times more likely to be diagnosed with autism than
  girls.
- Most children were still being diagnosed after age four, though
  autism can be reliably diagnosed as early as age two.
- Thirty-one percent of children with ASD have an intellectual
  disability (intelligence quotient [IQ] <70), 25 percent are in the
  borderline range (IQ 71–85), and 44 percent have IQ scores in the
  average to above average range (i.e., IQ >85).
- Autism affects all ethnic and socioeconomic groups.

What causes autism?

Those are the statistics, but what causes autism?

According to the National Institute of Health: "[I]n about 2 to 4 percent of people with ASD, rare gene mutations or chromosome abnormalities
are thought to be the cause of the condition."

So that accounts for up to 4 percent. What about the rest?

The National Institute of Health goes on to say: "Non-genetic factors
may contribute up to about 40 percent of ASD risk."

Forty percent plus 4 percent equals 44 percent. What about the other 56
percent? They seem to be unaccounted for.

What about the 40 percent that are non-genetic? If it's not genetic, then
it is environmental. So what environmental insult caused autism?

Some researchers have looked at vaccines.

The media would have you believe that the autism-vaccine connection
has been disproven. This is not true.

Here's what the media has not told you: there are many government scientists and reputable doctors who suspect there might be a connection between vaccines and autism and they have called for further
investigation.

Former head of the National Institutes of Health says the autism-vaccine link needs another look.

The former head of the National Institutes of Health, Dr. Bernadine Healey, said that the autism-vaccine link has *not* been "debunked." In 2008 she gave an interview and said:

> The more you delve into [the question if vaccines cause autism] . . . what I
> come away with is [that] the question has not been answered . . . We . . . have
> to . . . take another look at the hypothesis [that vaccines can cause autism] . . .
> not deny it . . . we have the tools today . . . to try [to] find out if . . . there is that
> susceptible group . . . Maybe there is a group of individuals . . . that shouldn't
> have a particular vaccine or shouldn't have vaccine on the same schedule . . .
> It is the job of the public health community and of physicians . . . to say, "Yes,
> we can make it safer."

Healey went on to comment about families who have reported that their child developed autism after getting vaccines:

> I think the government, or certain public health officials in the government,
> have been too quick to dismiss the concerns of these families without study-
> ing the population that got sick. I haven't seen major studies that focus on
> 300 kids who got autistic symptoms within a period of a few weeks [after
> getting] a vaccine. I think that the public health officials have been too quick
> to dismiss the hypothesis as irrational without sufficient studies of causation.
> I think they often have been too quick to dismiss studies in the animal labo-
> ratory of mice, of primates, that do show concerns with regard to vaccines and
> the mercury preservative in vaccines.

Healey then explained why the government does not want scientists to study this further and above all does not want to examine whether there is a susceptible group that should not keep to the recommended and/or mandated immunization schedule. She said:

> The government has said in a report by the Institute of Medicine—and by
> the way I'm a member of the Institute of Medicine; I love the Institute of
> Medicine—but in a report in 2004, it basically said, "Do not pursue suscep-
> tibility groups, don't look for those patients, those children who may be vul-
> nerable." I really take issue with that conclusion. The reason why they didn't
> want to look for those susceptibility groups was because they were afraid if

they found them, however big or small they were, that that would scare the public away.

Dr. Healy concluded her interview by wisely stating, "I don't think you should ever turn your back on any scientific hypothesis because you are afraid of what it might show."

The media wants you to believe that any doctor who thinks that there may be a vaccine-autism connection or that it should be looked at is a quack who preys on vulnerable parents.

Really?

Here's a little background on Dr. Bernadine Patricia Healy:

Dr. Bernadine Healy became the 13th NIH director in April 1991, appointed by President George H.W. Bush. Shortly after her appointment, she launched the NIH Women's Health Initiative, a $500 million effort to study the causes, prevention, and cures of diseases that affect women. She also established the Shannon Award, grants designed to foster creative, innovative approaches in biomedical research and keep talented scientists in a competitive system. Prior to her appointment, she was chairman of the Research Institute of the Cleveland Clinic Foundation, where she directed the research programs of nine departments including efforts in cardiovascular disease, neurobiology, immunology, cancer, artificial organs, and molecular biology. From her appointment in November 1985, she also served as a staff member of the clinic's department of cardiology.

Is she a quack?

Government admits the vaccine-autism connection in one case and pays.

The government agreed that vaccines triggered autism in a landmark case.

Hannah Poling was normal and happy in her first eighteen months of life. Her father, Jon, is a neurologist. Her mother, Terry, is an attorney and a nurse.

Then, in July 2000, she was vaccinated against nine diseases in one doctor's visit: measles, mumps, rubella, polio, varicella, diphtheria, pertussis, tetanus, and Haemophilus influenzae.

Afterward, her health declined fast. According to an article written about her: "She developed high fevers, stopped eating, didn't respond when spoken to, began showing signs of autism, and began having screaming fits."

In 2006, Jon Poling, along with three other researchers, all of whom were affiliated with Johns Hopkins at the time, published a case report and chart review regarding the association of mitochondrial disease and autism in the *Journal of Child Neurology*. Then, "In 2002, Hannah's parents filed an autism claim in federal vaccine court. Five years later, the government settled the case before trial and had it sealed."

Why would the government seal this case so no one else knows what happened? The facts did come out. It was revealed that Hannah Poling was awarded $1.5 million plus an additional $500,000 a year because vaccines aggravated an unknown mitochondrial disorder. Her life-long compensation is estimated to be twenty million dollars.

What is mitochondrial disease? According to the Cleveland Clinic: "Mitochondrial diseases are long-term, genetic, often inherited disorders that occur when mitochondria fail to produce enough energy for the body to function properly."

Can other children, like her, also be vulnerable? How many children in the United States have mitochondrial disease?

According to the Cleveland Clinic: "Each year, about 1,000 to 4,000 children in the United States are born with a mitochondrial disease."

If this disease makes children susceptible to getting autism from vaccines, shouldn't the parents be warned? Shouldn't there be testing so these children can avoid vaccines that can trigger autism?

The government says no. Vaccine mandates have no "carve out" for children with mitochondrial disorders.

The government has a one-size-fits-all mass vaccine program. Every child is treated as if they are genetically and biologically the same. But that's a lie because children are not all the same.

Dr. Frank DeStefano, the CDC head of vaccination, agreed that vaccines may trigger autism in susceptible children, but they are not studying it.

Perhaps no other person in the world is more responsible for denying the vaccine-autism link than Dr. Frank DeStefano.

DeStefano is the CDC's director of Immunization Safety. He's the top guy. It was his studies that supposedly did not find a link between the measles vaccines and autism.

When DeStefano was asked about the Hannah Poling case in an interview with reporter Sharyl Attkisson, his answer was shaky.

DeStefano was asked if other children with the same mitochondrial disease may be susceptible to autism, like Hannah, and whether these kids should be identified so they're not injured by vaccines.

DeStefano is unable to answer intelligently for a full twenty-five seconds. (You can hear him struggle on the recording of the interview.). Listen here: https://sharylattkisson.com/2018/12/cdc-possibility-that-vaccines -rarely-trigger-autism/#audio.

Here's the transcript of what this top vaccine doctor says:

> Uh, yeah, I mean, I think um . . . You know, I think it's something that, uh, well I mean, you know, in terms of uh . . . I mean, It's hard, it's hard to say, you know, I mean it's like, um . . . I mean how how important that is. I mean, it's a theoretical possibility, I guess the, the [Hanna] Poling case maybe suggested it could happen. Uh, but [unintelligible] cause it's hard to predict who those children might be, but certainly, um individual cases, uh, can be studied to try to, uh, to look at those, uh, those possibilities.

After stammering and stumbling for a painful twenty-five seconds, DeStefano admits that it's a "possibility" that vaccines can trigger autism, but "it's hard to predict who those children might be."

The problem is they are not even trying.

The public is being told that vaccines are safe for everyone and everyone should get them. Because this is the main message, reports that say anything else are suppressed.

Dr. William Thompson, CDC whistleblower, says information about the autism-vaccine link was not reported.

Who is William Thompson?

Dr. William Thompson is the co-author of the CDC study with Dr. Frank DeStefano that said there was no link between autism and vaccination. However, he has since declared:

> [The CDC is] not doing what they should be doing because they're afraid to look for things that might be associated . . .
>
> I was complicit, and I went along with this, we did not report significant findings [about autism and vaccines]. Ya know, I'm not proud of that and uh, it's probably, *it's the lowest point in my career that I went along with that paper.*

Here's what I shoulder. I shoulder that the CDC has put the research ten years behind. Because the CDC has not been transparent, we've missed ten years of research because the CDC is so paralyzed right now by anything related to autism. They're not doing what they should be doing because they're afraid to look for things that might be associated. . . . So that's the way I view all this. I am completely ashamed of what I did.

There is a biological, there is biologic plausibility right now, I really do believe there is, to say that thimerosal causes autism-like symptoms.

I've been involved in the separate situation unrelated to this, where these senior [CDC] people just do completely unethical, vile things and no one holds them accountable.

But I also have to say these drug companies and their promoters, they're making such a big deal of these measles outbreaks and they are now, they're making a big deal that polio is coming back and polio comes back all the time in third world countries. It's like a never-ending thing where the press loves to hype it and it scares people. It scares the crap out of people when they hype those two types of outbreaks. I think as they teach you at the CDC, you have to stay on message.

So here we have a CDC scientist who admits that the *"CDC is so paralyzed right now by anything related to autism. They're not doing what they should be doing because they're afraid to look for things that might be associated . . . "*

This statement is in the public record. It's all there for anyone who is interested, but the media does not report on it. Why is that?

Pilot study suggests vaccinated children are four times more likely to be diagnosed with autism versus non-vaccinated children.

There is also a Pilot Study from researchers at the School of Public Health, Jackson State University that looked at vaccinated and unvaccinated children in four states: Florida, Louisiana, Mississippi, and Oregon.

In this study, there were 405 vaccinated children and 261 unvaccinated children. The study was designed as a survey and diagnosis information was supplied by the children's mothers. They were asked to provide information based on vaccine records and diagnoses by a physician.

The survey found that nineteen out of 405 vaccinated children were diagnosed with Autism Spectrum Disorder (4.7 percent) and three out of 261 unvaccinated children were diagnosed (1.1 percent). Based on this limited pilot study there appears to be a much higher risk for autism among vaccinated children.

Vaccine proponents like to claim that this study was retracted. This is not true. It was never retracted. It is available online here: https://oatext.com/pdf/JTS-3-186.pdf.

In fact, other studies have looked at autism and vaccines in general and have found a relationship. For example, a study published in the *Journal of Toxicology and Environmental Health* was titled "A positive association found between autism prevalence and childhood vaccination uptake across the U.S. population." The study said:

> A positive and statistically significant relationship was found: The higher the proportion of children receiving recommended vaccinations, the higher was the prevalence of autism or speech or language impairment. A 1% increase in vaccination was associated with an additional 680 children having autism or speech or language impairment.

With all this information suggesting that there is an autism-vaccine connection, don't you think it's time for a thorough independent study by researchers who are not part of the vaccine industry?

Sources:

Attkisson, Sharyl. "CDC: "Possibility" That Vaccines Rarely Trigger Autism (AUDIO)." Sharyl Attkisson. Last modified December 13, 2018. https://sharylattkisson.com/2018/12/cdc-possibility-that-vaccines-rarely-trigger-autism/.

Attkisson, Sharyl. "Family to Receive $1.5M+ in First-Ever Vaccine-Autism Court Award." CBS News. Last modified September 10, 2010. https://www.cbsnews.com/news/family-to-receive-15m-plus-in-first-ever-vaccine-autism-court-award/.

"Autism Spectrum Disorder Fact Sheet." National Institute of Neurological Disorders and Stroke | National Institute of Neurological Disorders and Stroke. Last modified March 13, 2020. https://www.ninds.nih.gov/Disorders/Patient-Caregiver-Education/Fact-Sheets/Autism-Spectrum-Disorder-Fact-Sheet.

"Autism Spectrum Disorder." MedlinePlus: Genetics. Accessed February 11, 2021. https://ghr.nlm.nih.gov/condition/autism-spectrum-disorder#genes.

Barry, Kevin. *Vaccine Whistleblower: Exposing Autism Research Fraud at the CDC*. New York, NY: Simon & Schuster, 2015.

"Bernadine Healy." Wikipedia. Last modified December 15, 2005. https://en.wikipedia.org/wiki/Bernadine_Healy.

DeLong, Gayle. "A Positive Association found between Autism Prevalence and Childhood Vaccination uptake across the U.S. Population." *Journal of Toxicology and Environmental Health, Part A* 74, no. 14 (2011), 903–916.

"Former Head of National Institutes of Health on Vaccine-Autism Link." Sharyl Attkisson. Last modified January 6, 2019. https://sharylattkisson.com/2019/01/former-head-of-national-institutes-of-health-on-vaccine-autism-link/.

Mawson, Anthony R. "Pilot comparative study on the health of vaccinated and unvaccinated 6- to 12-year-old U.S. children." *Journal of Translational Science* 3, no. 3 (n.d.), 1–12.

"NIH Director Dr Bernadine Healy Speaks to Sharyl Attkisson about Autism Susceptibility." YouTube. n.d. https://www.youtube.com/watch?v=UZFPpHBNp2M.

Poling, Jon S., Richard E. Frye, John Shoffner, and Andrew W. Zimmerman. "Developmental Regression and Mitochondrial Dysfunction in a Child With Autism." *Journal of Child Neurology* 21, no. 2 (2006), 170–172.

Bernadine Healy, M.D. Director, National Institutes of Health, April 9, 1991–June 30, 1993 National Institute of Health https://www.nih.gov/about-nih/what-we-do/nih-almanac/bernadine-healy-md.

US Government Data Describes Children Becoming Autistic After Getting Vaccinated

> *Quick Version:* There are hundreds of cases in VAERS of children getting autism and autism symptoms after being vaccinated.

Does brushing one's teeth cause accidents?

Let's say you lived in the same house for twenty years and one day you were brushing your teeth when a car hit a fire hydrant outside your home. It only happened once in twenty years. Does that mean that brushing your teeth causes cars to hit fire hydrants? Probably not.

Just because something happens when another thing happens doesn't mean that one thing causes the other.

This is the difference between a coincidence and causal relationship (x causes y).

People whose job it is to protect the vaccine industry are very quick to make analogies like this. They tell us that when perfectly healthy children who are developing normally and hitting all their milestones suddenly become autistic after getting vaccines, it means nothing. It's as valid a connection as teeth brushing causing accidents.

They say that it's only a coincidence.

Proving that something is not a coincidence and that there is a causal relationship takes research. The problem is that no one is doing the independent research into individual cases when a child suddenly becomes autistic after vaccination. We are told that CDC-funded studies conveniently prove that vaccines don't cause autism. Case closed. No more investigation needed.

We are also told, however, that 40 percent of the time autism is caused by the environment. We have evidence of children radically changing after vaccination. We have evidence of money paid to a child by the government whose vaccine caused her to become autistic. We have statements of government scientists saying it's a "possibility" that vaccines can trigger autism. We have a pilot study showing that vaccines increase the risk of autism by four times, and yet the independent research into individual cases is not being done.

Examples of autism following vaccines

What we are left with is data like this from the government's own database (VAERS). Since no research is being done into any of these cases, you can decide for yourself. Did vaccines cause autism or is it just a coincidence?

Is this section we are looking at childhood vaccines in general, not just the measles vaccine. The VAERS ID is provided so you can look up these cases yourself. These are just a handful of examples; there are many, many more in the database.

<u>One-year-old boy in Washington</u>
Two days after vaccination developed somnolence, fever, rash, constant moaning. Before vaccination using 20+ words fluently, since 2/16/17 he is non-verbal. He has been diagnosed with speech regression, developmental delay, encephalopathy, autism spectrum disorder.
VAERS ID # 733487–1

<u>One-year-old boy in Michigan</u>
He got cellulitis, and severe diarrhea. He also started to regress and started showing signs of autism. His Dr is in agreement that all this was caused from his vaccines at 1 year of age.
VAERS ID # 735581–1

Fourteen-month-old boy in California
Swelling and rash were immediate, then fever for 3 days then child was never the same, no eye contact, moderate to severe autism diagnosis followed, recently test results have shown off the chart high aluminum poisoning. VAERS ID # 774170–1

One-year-old boy (State not provided)
Fever, rash covering body, trouble waking him up, swollen and painful injection site. Within a week he was unable to walk and was in therapy until age 2. He did not walk again until 21 months. He stopped eating solid and was also in therapy to help as well with feeding. He stopped talking and was diagnosed with autism a month later. VAERS ID # 800239–1

One-year-old girl in Mississippi
Legs began to swell with large knots on both legs at the injection site, lasting a few weeks. She began running a fever which lasted about a week. She would no longer speak, hold eye contact, clap, wave, give hugs and kisses, play with her siblings, etc. She would just sit and stare or play alone. She would no longer respond to her name or any sounds to get her attention. She was tested on August 2017 and in less than 10 minutes they were able to diagnose her with autism spectrum disorder. She has been in therapy over a year now and is still non-verbal. VAERS ID # 800384–1

One-year-old girl in Washington
Fever, bright red cheeks, measles type rash on legs, after fever stopped responding to name, stopped eye contact, stopped all nonverbal communication, stopped all verbal communication. VAERS ID # 801883

US Government Pays Families After Child Gets Encephalitis from Vaccines (But the Description of Encephalitis Sounds Like Autism)

> *Quick Version:* The government has paid many millions of dollars to families whose children got encephalitis immediately after vaccines. However, the symptoms for many of these kids sound a lot like autism.

The government denies there is an autism-vaccine link. However, it does not deny and pays money when there is a link between vaccines and encephalitis.

Encephalitis is inflammation of the brain, and it can be caused by any one of a variety of viruses.

Encephalitis is a different diagnosis than autism, but when you read the descriptions of injured children they sound strangely alike. Here are some examples:

Ryan Mojabi: autism or encephalitis after vaccines

In this case, the government paid $989,955.25 plus an annuity to a child named Ryan Mojabi. Ryan became autistic after his MMR vaccine. First the claim said the child got autism. Then it was changed to say the child got encephalitis. The symptoms were the same, but apparently, it's easier for the government to admit that encephalitis was caused by the vaccine, not autism.

Here's what the legal decision said:

> [It was] alleged that as a result of "all the vaccinations administered to [Ryan] from March 25, 2003, through February 22, 2005, and more specifically, measles-mumps-rubella (MMR) vaccinations administered to him on December 19, 2003 and May 10, 2004," Ryan suffered "a severe and debilitating injury to his brain, described as Autism Spectrum Disorder ('ASD')."

But a supplemental filing emphasized encephalitis.

The document says:

> On June 9, 2011, respondent filed a supplemental report pursuant to Vaccine Rule 4(c) stating . . . that Ryan suffered a Table injury under the Vaccine Act—namely, an encephalitis within five to fifteen days following receipt of the December 19, 2003 MMR vaccine . . . and that this case is appropriate for compensation under the terms of the Vaccine Program.

Get autism after vaccination and the government makes you play this stupid game of calling it encephalitis.

Encephalitis after vaccination

Here are other reports in VAERS where a child is injured by a vaccination. The description "encephalitis" sounds a lot like autism:

Fourteen-Month-Old Boy in Ohio

This boy became completely nonverbal, engaged in repetitive behavior and struggled with social skills after vaccines. They called it encephalitis. (Note that "pt" is an abbreviation for patient.)

> Leading up to the date of 6-18-2015, patient had several episodes of ear infections treated with Amoxicillin. On 6/18, his pediatrician at that time injected pt with 8 combination vaccines (listed above) while pt still had an active ear infection.
>
> After this, pt slowly lost his ability to verbally communicate. Before the vaccines, pt had a vocabulary of around 50 words and after he is only able to verbalize about 5 words. By the beginning of 2016, pt lost verbal communication and social skills.

He began stimming, spinning in circles and engaging in repetitive behavior. Today, he is still almost completely nonverbal and struggles with social skills. He was diagnosed by Hospital with Encephalopathy and by Dr. with Encephalitis.
VAERS ID # 744781–1

One-Year-Old Boy in Wisconsin
In this example, a boy "lost eye contact, speech and responsiveness" after vaccination. They called it encephalitis.

My son lost eye contact, speech and responsiveness to those around him following his 1 year vaccinations. MMR, varicella, pneumococcal. He has immune deficiency and encephalitis.
VAERS ID # 726833–1

One-Year-Old Boy
In this example, a child is described as possibly having both encephalitis and autism:

Child immediately developed fever, rash and discomfort. Within 2 weeks behavior changed dramatically, language skills regressed, frequent fevers, seizures, loss of eye contact. 8 months out diagnosed with autism spectrum disorder. Has been diagnosed with Encephalitis. 3/4/09 MR received from PCP from 2/5/07 to 10/2008. DX: Autism Spectrum Disorder. Seen 10/19/07 with parental concerns re: speech development. No real progress since age 1. Does not respond to name when called. Motor development WNL. More concerns with behavior reported 11/29/07 along with frequent fevers. Has episodes of stiffening and unresponsiveness and waking screaming. DX: Autism Spectrum D/O 12/2007. VAERS ID # 340268–1

Sources:
Vaccine Adverse Events Reporting System (VAERS) CDC https://wonder.cdc.gov/vaers.html.
"Office of Special Masters Decision Awarding Damages." US Court of Federal Claims. Accessed February 11, 2021. https://www.uscfc.uscourts.gov/sites/default/files/opinions/CAMPBELL-SMITH.MOJABI-PROFFER.12.13.2012.pdf.

Studies Have Found a Biological MMR Vaccine-Autism Connection

> *Quick Version:* Research has discovered that the measles vaccine can cause an auto-immune reaction in some children where the child's immune system attacks parts of the child's brain cells.

A few studies have looked at what the measles vaccine does in the brain and how the brains of autistic children are different from non-autistic children. What they have found is disturbing.

Apparently, an autoimmune reaction to the measles *vaccine* can destroy parts of the brain.

The measles vaccine can cause an auto-immune reaction in some children where the child's immune system attacks some of the child's own brain cells.

One study compared 125 autistic children and ninety-two children who did not have autism. The study found that autistic children had higher levels of measles antibodies.

In fact, over 90 percent of the autistic children with measles antibodies were also positive for antibodies against myelin basic protein or MBP. The non-autistic children did not have this.

Why is this important?

Myelin basic protein (MBP) is what makes myelin and myelin is what covers and protects nerve cells in the brain and nervous system.

If the measles vaccine is making some children attack their own myelin, then the nerve cells in their brains cannot do what they are supposed to. If nerve cells cannot do what they are supposed to, then the child's brain doesn't function as it should.

In this study, the scientists concluded, "Stemming from this evidence, we suggest that an inappropriate antibody response to MMR, specifically the measles component thereof, might be related to pathogenesis of autism."

This is a very important discovery, so why has it been ignored?

If certain children develop an autoimmune response to the measles vaccine and that response attacks parts of their brain, don't you think this is important?

According to this study, this auto-immune response only happens in children with autism. Or is this also just a coincidence?

Now we have a possible biological mechanism that explains how the measles vaccine can cause autism in some children, yet the government ignores it. Why?

Sources:

"Catalytic Autoantibodies Against Myelin Basic Protein (MBP) Isolated from Serum of Autistic Children Impair In Vitro Models of Synaptic Plasticity in Rat Hippocampus." *Journal of Neuroimmunology*, October 2015, 287. doi: 10.1016/j.jneuroim.2015.07.006.

Singh, V. K. "Abnormal Measles-Mumps-Rubella Antibodies and CNS Autoimmunity in Children with Autism." *Journal of Biomedical Science* 9, no. 4 (July/August 2002), 359–364.

SECRET #28

The Vaccine-Autism Connection Is Being Ignored and Denied Because It Would Cost the US Government Trillions of Dollars

---

Quick Version: If the health authorities accepted the fact that vaccines can cause autism in some children, it could literally bankrupt the federal government.

Does the measles vaccine or any vaccine cause autism?

Many parents have seen their normal child who was meeting all their milestones become autistic within days of getting their shots.

But this is merely parent information. Anecdotal accounts. The authorities tell us that this information cannot be trusted and doesn't count.

Every talking head on TV and internet troll tells us that it is totally 100 percent certain that vaccines can never cause autism.

Is that the truth?

Summary of a fraction of the evidence that autism and vaccines are connected for some children

Here are some facts:

- The former head of the National Institute of Health says that we have to take another look at the hypothesis that vaccines can cause autism, not deny it.
- The government has paid about twenty million dollars to a child who got autism after vaccines (Hannah Polling).
- The government has also paid many more millions of dollars to other families whose children got encephalitis after vaccines. Interestingly, the symptoms for many of these kids sound a lot like autism.
- Frank DeStefano, the CDC's director of Immunization Safety, says it's hard to predict which children might be susceptible to autism from vaccines and they are not studying that issue.
- CDC whistleblower Dr. William Thompson said that the CDC did not report significant findings about autism and vaccines and has put the research ten years behind "because the CDC is so paralyzed right now by anything related to autism."
- A pilot study found that nineteen out of 405 vaccinated children were diagnosed with Autism Spectrum Disorder (4.7 percent) and three out of 261 unvaccinated children were diagnosed (1.1 percent).
- There are hundreds of cases in VAERS of children getting autism-like symptoms after being vaccinated.
- Research has discovered that the measles vaccine can cause an autoimmune reaction in some children where the child's immune system attacks parts of the child's own brain cells.

Has the media shared any of this information? No, they haven't. Why is that?

Imagine if the government accepted the fact that vaccines can cause autism in some children. This is what would happen:

- The Federal Government would have to pay out trillions of dollars.
- The country and the world would lose faith in the vaccine industry. The vaccine/pharmaceutical industry is one of the main advertisers on TV. The media depends on the revenue it receives from Big Pharma.
- Heads would roll at the CDC, the FDA, and the Department of Health and Human Services and lots of powerful people would lose their jobs.

Would it really cost trillions of dollars? Yes. Here's why:

The one case where the Federal Government paid a family for the vaccine causing autism (the Hannah Polling case discussed earlier) resulted in a settlement of approximately twenty million dollars. This was money to take care of this vaccine-injured child for the rest of her life.

In 2018, the United States' Health Resources and Services Administration (HRSA) "found that about 1.5 million U.S. children—or 1 in 40—have received a diagnosis of, and currently have, autism spectrum disorder."

The National Institute of Health says that: "Scientists believe that both genetics and environment likely play a role in ASD [autism]." What is it in the environment? If it's vaccines, then that would affect 300,000 children per year. Multiply 300,000 by twenty million dollars per child. That's six trillion dollars.

That's a lot of money. It's almost double the total amount of money the US Government gets in revenue in an entire year.

So, if the government accepted the fact that the vaccine-autism connection was real, almost every dollar the United States gets in revenue for two full years would be needed just to pay families whose children were injured by the vaccines.

That's six trillion reasons to say that there is absolutely no connection between autism and the vaccines.

Sources:

Attkisson, Sharyl. "CDC: 'Possibility' That Vaccines Rarely Trigger Autism (AUDIO)." Sharyl Attkisson. Last modified December 13, 2018. https://sharylattkisson.com/2018/12 /cdc-possibility-that-vaccines-rarely-trigger-autism/.

Barry, Kevin. *Vaccine Whistleblower: Exposing Autism Research Fraud at the CDC*. New York, NY: Simon & Schuster, 2015.

"HRSA-led Study Estimates 1 in 40 U.S. Children Has Diagnosed Autism." U.S. Health Resources & Services Administration. Last modified November 26, 2018. https://www.hrsa.gov/about/news/press-releases/hrsa-led-study-estimates-children -diagnosed-autism.

Mawson, Anthony R. "Pilot comparative study on the health of vaccinated and unvaccinated 6- to 12-year-old U.S. children." *Journal of Translational Science* 3, no. 3 (n.d.), 1–12.

"NIH Director Dr Bernadine Healy Speaks to Sharyl Attkisson about Autism Susceptibility." YouTube. n.d. https://www.youtube.com/watch?v=UZFPpHBNp2M.

Poling, Jon S., Richard E. Frye, John Shoffner, and Andrew W. Zimmerman. "Developmental Regression and Mitochondrial Dysfunction in a Child With Autism." *Journal of Child Neurology* 21, no. 2 (2006), 170–172.

National Institute of Health Autism Spectrum Disorder Fact Sheet. https://www.ninds
.nih.gov/Disorders/Patient-Caregiver-Education/Fact-Sheets/Autism-Spectrum
-Disorder-Fact-Sheet#3082_5.

PART 6

Vaccine Companies Are Not Held Accountable When a Child Is Hurt or Killed by a Vaccine

Drug companies like Merck do not pay when one of their childhood vaccines hurts or kills a child. We do through the tax on every vaccine. Do you think that's fair when these corporations make billions of dollars a year from their vaccines? It is all reward, no risk for them. But, we, the consumers, are left holding the bag by paying more money to compensate families when children are vaccine-injured or sacrificing the health of our own children.

Secrets:
29. Thanks to the US Government, Vaccine Makers Are Not Responsible When Their Vaccine Hurts or Kills Children
30. The Government Has Already Paid Almost Four Billion Dollars to People Killed or Injured by Vaccines, But Getting Compensated Is Not Easy

Thanks to the US Government, Vaccine Makers Are Not Responsible When Their Vaccine Hurts or Kills Children

> *Quick Version:* In 1986, the government set up the National Vaccine Injury Compensation Program (VICP). Now, when someone is hurt or killed by a vaccine, the manufacturer doesn't have to pay, we do. Because of this, vaccine makers have little incentive to make their vaccines safer.

Merck makes the measles vaccines. In 1986, Merck and other vaccine manufacturers urged Congress to pass the National Childhood Vaccine Injury Act.

Drug companies argued that their products were so risky and "unavoidably unsafe" that they had to be protected from liability or they would stop making the vaccines.

A law protects vaccine makers when their vaccine hurts or kills a child.
The law was passed. It protects pharmaceutical companies from having to pay if their vaccines hurt or kill someone.

This is how it works:

The government set up something called the "National Vaccine Injury Compensation Program" (VICP). The VICP has a fund to pay people who are hurt or killed by vaccines.

The fund gets its money from a $.75 tax on all CDC recommended vaccines. For example, the measles-mumps-rubella vaccine is taxed at $2.25 per dose because there are three vaccines in it. All this money goes into a fund.

This means that when someone is killed or injured by a vaccine, the vaccine manufacturer does NOT have to pay. We do!

Injured people cannot sue the vaccine maker.

In fact, vaccine makers can't even be sued for defective manufacture, inadequate directions or warnings, and defective design. Do you think that helps the vaccine makers focus on safety?

Remember, in the United States, the vaccine companies have one primary customer—the US Government.

When the government approves a vaccine and adds it to the childhood vaccine immunization schedule, it means billions of dollars flow into the pockets of that drug company.

The process is designed to protect the vaccine maker, not the injured child.

When the National Childhood Vaccine Injury Act was interpreted by the US Supreme Court in the case of *Bruesewitz v. Wyeth*, it helped vaccine makers even more.

Here's the upshot:

- When a child is injured or killed by a vaccine, you have to go through vaccine court (Federal Court of Claims).
- You are not allowed "discovery." This means you are not allowed to ask for documents from the manufacturer like you can in a normal lawsuit.
- There is no judge or jury. There is only a "Special Master" who works for the government.
- If you lose in vaccine court you can technically sue in regular court, but you really can't because the US Supreme Court said that all product liability claims are barred. In other words, when a product harms or kills someone, you can normally claim that the product

was made wrong or labeled wrong or designed wrong. But, the US Supreme Court said you *cannot* do that with vaccines.

- Two Supreme Court justices (Justice Sonia Sotomayor and the late Justice Ruth Bader Ginsburg) disagreed with majority opinion. They recognized that this terrible law gives vaccine makers no reason to make their vaccines safer. They understood that with this unprecedented legal protection and profits already rolling in that manufacturers had no incentives to improve their vaccines and make them safer. In their dissent opinion in the *Bruesewitz* case, they wrote:

  > [T]he majority's decision leaves a regulatory vacuum in which no one—neither the FDA nor any other federal agency, nor state and federal juries—ensures that vaccine manufacturers adequately take account of scientific and technological advancements. This concern is especially acute with respect to vaccines that have already been released and marketed to the public. Manufacturers, given the lack of robust competition in the vaccine market, will often have little or no incentive to improve the designs of vaccines that are already generating significant profit margins.

- In other words, these Supreme Court justices are saying that the present system gives companies like Merck no reason to make safer vaccines. Why? Because they have no competition, no liability, and they are already making lots of money from their vaccine that's on the market.

The vaccine model is designed to protect industry profits, not children's lives.

So here's the money making business model as it exists in the United States:

- Drug companies like Merck have one major vaccine customer, the US Government.
- This customer *mandates (forces)* consumers (children) to use their product through the CDC and the states.
- The product is paid for by US Government or you.
- When someone is injured by the product, the manufacturer doesn't have to pay, you do.

This is a big reason why there are approximately seventy vaccines on the childhood vaccine schedule today compared to about twenty a generation ago.

This is also a big reason why the World Health Organization says there are 120 new vaccines in the pipeline.

This is also a big reason why the benefits of vaccines are exaggerated and the actual risks and side effects are minimized or ignored.

If drug companies actually had to pay when their vaccine hurt or killed a child, do you think they would act the same way? Or would they be a lot more careful in testing vaccines and only offer those that were proven safe and absolutely necessary?

Sources:

"$4 Billion and Growing: U.S. Payouts for Vaccine Injuries and Deaths Keep Climbing." Children's Health Defense. Last modified November 19, 2018. https://childrenshealth defense.org/news/4-billion-and-growing-u-s-payouts-for-vaccine-injuries-and-deaths-keep -climbing/.

"2018 NVIC Position on National Childhood Vaccine Injury Act of 1986—NVIC." National Vaccine Information Center (NVIC). Last modified May 2018. https: //www.nvic.org/injury-compensation/nvic-position-on-1986-childhood-vaccine-injury -act.aspx.

"BRUESEWITZ ET AL. v. WYETH LLC." *Supreme Court of the United States.* February 22, 2011. https://www.supremecourt.gov/opinions/10pdf/09–152.pdf.

Kaddar, Miloud. "Global Vaccine Market." World Health Organization. Accessed February 12, 2021. https://www.who.int/influenza_vaccines_plan/resources/session_10 _kaddar.pdf.

"National Vaccine Injury Compensation Program." U.S. Health Resources & Services Administration. Last modified July 23, 2020. https://www.hrsa.gov/vaccine-compensation /index.html.

SECRET #30

The Government Has Already Paid Almost Four Billion Dollars to People Killed or Injured by Vaccines, But Getting Compensated Is Not Easy

Quick Version: About four billion dollars has already been paid to about six thousand children injured or killed by vaccines. Getting justice through the vaccine courts is a difficult, long, and painful process.

The National Vaccine Injury Compensation Program (NVICP) fund has paid out almost four billion dollars to nearly six thousand vaccine victims.

Remember, when someone is injured or killed by a vaccine, the money to compensate comes from an excise tax that's charged on every vaccine. The manufacturer of the vaccine does not pay.

If vaccines are safe, then why would the US Government pay
almost six thousand people a total of four billion dollars for
their vaccine deaths or injuries?

Year	Total Payments to Vaccine Injured
1989	$1,371,761.92
1990	$54,689,215.73
1991	$98,842,061.28
1992	$98,752,676.41
1993	$125,402,993.98
1994	$104,889,607.13
1995	$110,014,172.61
1996	$105,885,679.89
1997	$119,397,874.59
1998	$133,484,353.24
1999	$101,024,548.76
2000	$131,782,015.74
2001	$111,318,723.12
2002	$63,109,448.07
2003	$87,509,650.06
2004	$66,211,708.71
2005	$59,551,048.33
2006	$52,540,994.37
2007	$97,175,608.51
2008	$83,556,982.40
2009	$85,345,704.64
2010	$189,261,439.67
2011	$233,482,659.44
2012	$186,803,360.70
2013	$276,676,893.56
2014	$223,563,646.17
2015	$225,219,939.33
2016	$252,610,672.33
2017	$282,096,906.34
2018	$226,628,298.86
2019	$225,457,657.94
2020 (as of January)	$57,255,770.19

Although you just read that the government has paid over four billion dollars in damages related to vaccines, this compensation is hard won. It is a long, difficult road for parents to walk in order to pursue a lawsuit. Many don't succeed. Many give up along the way or even do not have the energy to begin.

Imagine your child was injured or killed by a vaccine. You want whatever happened to your child to not happen to other children. You want to hold the company that caused the injury responsible for what they did. You want help to pay for all the new medical bills that come with a severe vaccine injury. You want some form of justice.

You can't sue the vaccine maker, but you can file a claim with the Federal Court of Claims (the vaccine court set up by the National Vaccine Compensation Program). Sounds easy? It's not.

Three stories that illustrate how compensation from the vaccine court works

Here are three stories reported by the *Los Angeles Times* about children who were injured or killed by vaccines. Their families had to fight for many years to receive justice:

Veronica Spohn

The parents of Veronica Spohn believed that a DPT shot caused their infant daughter to suffer brain damage. They lost in the United States Court of Federal Claims on a technicality because their petition was filed a few hours late. Veronica's mother, Karen Spohn, a nurse said, "I had a normal child, and all of a sudden in one day, within hours of the vaccine, she became a child with a disability who is going to need assistance for the rest of her life. . . . They didn't rule that she didn't have damage. All they did was say, you filed twelve hours too late—too bad on you."

Dustin Barton

Dustin suffered seizures and brain damage after getting a DPT shot. His mother, Lori Barton, filed his claim in November 1991, but the case dragged on for years. Barton believed that the government was waiting for Dustin to die because it would be cheaper to pay the death benefit of $250,000 than to pay for an annuity to cover lifetime care. In 1997, Dustin died of a seizure, but the government continued to fight. In May 2000, eight and a half years after their petition was filed, the government admitted that Dustin's injuries were vaccine-related.

Rachel Zuhlke

Rachel Zuhlke also was brain injured after getting the DPT shot. Her family filed a claim in September 1992, but the government blamed her brain injuries on complications from a strep infection. Nine years later, in July 2001, the government ruled that Rachel was entitled to compensation. Although relieved that the case was finally over, Rachel's mother is still saddened by what happened to her child, now a young woman. Rachel's life, she said, "is so different from what it should be at twenty."

Sources:

"$4 Billion and Growing: U.S. Payouts for Vaccine Injuries and Deaths Keep Climbing." Children's Health Defense. Last modified November 19, 2018. https://childrenshealth defense.org/news/4-billion-and-growing-u-s-payouts-for-vaccine-injuries-and-deaths -keep-climbing/.

"2018 NVIC Position on National Childhood Vaccine Injury Act of 1986." National Vaccine Information Center (NVIC). Last modified May 2018. https://www.nvic .org/injury-compensation/nvic-Position-on-1986-childhood-vaccine-injury-act.aspx.

Levin, Myron. "Vaccine Injury Claims Face Grueling Fight. Victims Increasingly View U.S. Compensation Program as Adversarial and Tightfisted." Los Angeles Times. Last modified November 29, 2004. https://articles.latimes.com/2004/nov/29/business /fi-vaccinecourt29.

"National Vaccine Injury Compensation Program." U.S. Health Resources & Services Administration. Last modified July 23, 2020. https://www.hrsa.gov/vaccine-compensation /index.html.

Singh, V. K. "Abnormal Measles-Mumps-Rubella Antibodies and CNS Autoimmunity in Children with Autism." *Journal of Biomedical Science* 9, no. 4 (July/August 2002), 359–364.

"Vaccine Compensation Data and Statistics." U.S. Health Resources & Services Administration. Accessed February 12, 2021. https://www.hrsa.gov/sites/default/files /hrsa/vaccine-compensation/data/data-statistics-vicp.pdf.

PART 7

The Government Is Forcing a One-Size-Fits-All Healthcare Plan That Does Not Fit All

The US vaccine program is based on two assumptions that are both wrong. The first assumption is that vaccines work all the time. This is not true. The second is that vaccines are safe for every child. That is also not true. In fact, both statements are untrue for the same reason. They both assume that all children are biologically and genetically identical and vaccines will act the same in every child. This is false. It is false with prescription drugs and it is false with vaccines. It is also the reason that the idea of a safe and effective one-size-fits-all mass vaccine campaign is wrong.

Secrets:
31. **Vaccinating Your Child to Protect Your Neighbor Is Not Scientifically Sound**
32. **Herd Immunity Is an Illusion**
33. **The Rates of Contagious Diseases Have Come Down Because of Better Public Health**
34. **The Measles Vaccine Can Actually Cause Measles**
35. **One-Size-Fits-All Vaccine Programs Do Not Work and Are Unsafe for Children Because (Surprise) All Children Are Not All the Same**

SECRET #31

Vaccinating Your Child to Protect Your Neighbor Is Not Scientifically Sound

> *Quick Version:* The media and the vaccine makers use guilt to push their products. They tell us that if we don't vaccinate our kids we harm our neighbors. This is false because vaccines don't always work, so your neighbor can still get measles.

Vaccine proponents use guilt and shame to try to make people vaccinate their children. We are told that we have to protect other children by vaccinating our own children. We have to protect our neighbor.

That is a very compelling idea, but it is not based on science.

1. If vaccines work all the time, then if your neighbor is vaccinated, they are protected regardless of what you do.
2. If vaccines don't work all the time then it doesn't matter if you are vaccinated or your neighbor is vaccinated; they could still get a disease like measles.

It turns out, number two is correct.

Scientists who study infections are called epidemiologists. Epidemiologists have admitted in their scientific publications that the measles vaccine is not 100 percent effective. Here are some examples:

In a study in Finland, researchers found "153 cases of measles among vaccinated individuals." They concluded that it's "more common than suggested" that the vaccine fails to work.

The scientists said: "In the single isolated rural community that experienced an explosive school-based [measles] outbreak, even the re-vaccinees [people vaccinated more than once] had a high measles risk . . . vaccine failures are probably more common than suggested."

Here are other examples:

- In the Netherlands, a study found eight people with measles. Six had been vaccinated twice. One was vaccinated once. One was unvaccinated.
- A 2014 study in New York found that a person vaccinated twice (a twenty-two-year-old woman) spread measles to four other people and two of the four were vaccinated.
- In March 2003, in Pennsylvania, a measles "outbreak" occurred in a boarding school in Pennsylvania with over six hundred students. There were nine cases of measles. Six of the nine had received two doses of measles vaccine prior to the outbreak. One had received one dose. Two of the nine were not vaccinated.
- Another "outbreak" of measles occurred in Waltham, Massachusetts in 1984. The scientists wrote that "This outbreak of measles occurred in a highly vaccinated population and 70% of the cases were vaccine failures." This means that 70 percent of the kids who got measles had been previously vaccinated.
- In another example, a measles outbreak occurred in what scientists called a "Highly Vaccinated Population." Of 122 patients with confirmed measles, ten were previously vaccinated against measles.
- Let's look at the most recent data from 2019. According to the CDC's Morbidity and Mortality Weekly Report Early Release there were 1,249 reports of measles cases in 2019. The CDC states: 142 of the 1,249 measles cases were in people *who were vaccinated* and 235 of the 1,249 of the measles cases may have been vaccinated because their vaccination status was unknown.

The fact that the measles vaccine is not fool proof is well known among people who make their living from studying or promoting vaccination.

Walter A. Orenstein was the former director of the United States' National Immunization Program, a post he held from May 1993 to January 2004. In a paper published in 1986, he and his co-authors wrote, "However, since the [measles vaccine] is not 100 percent effective, there will always be some proportion of vaccines (10 percent or less) who remain susceptible to the disease."

Remember this statement when the media and the government blame unvaccinated people for any "outbreak." The doctors know that fully vaccinated children can also get measles. It is unfair and misleading to suggest otherwise.

It also means that vaccinating your child in order to protect your neighbor is not scientifically sound. It may work or it may not because, as the scientists say, the measles vaccine is "not 100 percent effective."

Sources:

Edmonson, M. B. "Mild Measles and Secondary Vaccine Failure During a Sustained Outbreak in a Highly Vaccinated Population." *JAMA: The Journal of the American Medical Association* 263, no. 18 (1990), 2467–2471.

Hahné, Susan J., Laura M. Nic Lochlainn, Nathalie D. Van Burgel, Jeroen Kerkhof, Jussi Sane, Kioe B. Yap, and Rob S. Van Binnendijk. "Measles Outbreak Among Previously Immunized Healthcare Workers, the Netherlands, 2014." *Journal of Infectious Diseases* 214, no. 12 (December 2016), 1980–1986.

Nkowane, B. M., S. W. Bart, W. A. Orenstein, and M. Baltier. "Measles outbreak in a vaccinated school population: epidemiology, chains of transmission and the role of vaccine failures." *American Journal of Public Health* 77, no. 4 (April 1987), 434–438. doi:10.2105/ajph.77.4.434.

Paunio, M., K. Hedman, and I. Davidkin. "Secondary measles vaccine failures identified by measurement of IgG avidity: high occurrence among teenagers vaccinated at a young age." *Epidemiology and Infection* 124, no. 2 (April 2000), 263–271. doi:10.1017 /s0950268899003222.

Rosen, J. B., J. S. Rota, C. J. Hickman, S. B. Sowers, S. Mercader, P. A. Rota, W. J. Bellini, et al. "Outbreak of Measles Among Persons With Prior Evidence of Immunity, New York City, 2011." *Clinical Infectious Diseases* 58, no. 9 (May 2014), 1205–1210.

Yeung, L. F. "A Limited Measles Outbreak in a Highly Vaccinated US Boarding School." *Pediatrics* 116, no. 6 (December 2005), 1287–1291.

SECRET #32

Herd Immunity Is an Illusion

> *Quick Version:* We are told that somewhere between 92 percent and 95 percent of a population has to be vaccinated to provide "herd immunity" which will stop epidemics. This is an illusion because the percentages of vaccinated individuals never reach that high and yet the epidemics do not exist.

Have you heard of "Herd Immunity"? Sometimes it is called "Community Immunity." This is the idea that a lot of people must be immune to a particular disease so the disease can't spread.

The American Academy of Pediatrics defines it this way:

> In theory, herd immunity means not everyone in a community needs to be immune to prevent spread of disease. If a high enough proportion of individuals in a population are immune, the majority will protect the few susceptible people because the pathogen is less likely to find a susceptible person.

The CDC says this:

> To maintain herd immunity, communities have to vaccinate enough residents to protect the small number of people who cannot receive a vaccination for medical reasons. For example, medical experts say that between 92% and

95% of children should receive two doses of the measles, mumps, and rubella
(MMR) vaccine to maintain herd immunity against measles.

Herd immunity is a myth because vaccine rates are way below their magical numbers of 92–95 percent.

We are being told that if there is no herd immunity, then diseases will
spread. It sounds logical, but it's a myth because the vaccination rate for
every so-called "vaccine preventable disease" is well below 92 percent, but
there are no actual epidemics!

Here are the actual rates of vaccination for adults as reported by the
CDC from 2010–2016:

Estimated proportion of adults ≥19 years who received selected vaccines, by age group and increased risk status— National Health Interview Survey, United States, 2010–2016.

Vaccine	Average Percentage of Adults Who Received It
Influenza—age ≥19 yrs	41%
Pneumococcal—age 19–64 yrs, increased risk	21%
Pneumococcal—age ≥65 yrs	62%
Tetanus-toxoid (Td or Tdap)—age ≥19 yrs	62%
Tdap—age 19–64 yrs	18%
Tdap—age ≥65 yrs	14%
Hepatitis A—age ≥19 yrs	9%
Hepatitis B—age ≥19 yrs	26%
Herpes zoster—age ≥60 yrs	24%
HPV females—19–26 yrs	36%
HPV males- 19–26 yrs	7%

Every single disease on the list has a vaccination rate under 63 percent.
That's far below the magical 92–95 percent number for herd immunity.
Where are all the outbreaks? There aren't any.

So how do we get "herd immunity" if the adults aren't cooperating?

Think about schools. Children spend thousands of hours a year in rooms
with teachers and administrators. Are these adults immune to measles? The
janitors? The coaches? Are they all fully vaccinated? All got their boosters for
every vaccine? Of course not.

When you add in all the adults, chances are the total number of vaccine recipients gets well below 95 percent. There should be outbreaks in every school, but there aren't.

Today the CDC says:

> Measles is highly contagious, so anyone who is not protected against measles is at risk of getting the disease. People who are unvaccinated for any reason, including those who delay or refuse vaccination, risk getting infected with measles and spreading it to others.

But, before the measles vaccine became an advertising slogan, the CDC sang a different tune.

In January 2001, the CDC published a paper called "National serologic survey of measles immunity among persons 6 years of age or older, 1988–1994."

In that paper, the CDC said, **"Nearly all persons (99%) born in the pre-vaccine era (before 1957) were immune."**

Think about that for a moment. The CDC is telling us that 99 percent of people born before the measles vaccine even existed were immune from measles. So why exactly did we need a vaccine?

In that same paper, the CDC says, "Immunity declined among persons born in the vaccine era (after 1956) to 81% among those born in 1967–1976, and increased again to 89% among those born in 1977–1988."

So this means:

1. People born before the measles vaccine had more immunity to measles.
2. People today between the ages of thirty-four to forty-three have an 81 percent immunity rate. That's also far below the magical 92 percent–95 herd immunity rate, so why aren't there outbreaks of measles every day among these adults?

Air Force measles vaccine rates in the low eighties but no "epidemics."

Here's another study that is very revealing. The Air Force wanted to know how many of its new recruits were immune to measles, mumps, and rubella. Blood was tested from 32,502 recruits from Joint Base San Antonio-Lackland from April 25, 2013 to April 24, 2014. Here are the results:

Disease	Sero-prevalence
Measles	81.6%
Mumps	80.3%
Rubella	82.1%

All were far below the 92–95 percent level needed for "herd immunity." Have you heard about the outbreaks or "epidemics" of measles, mumps, and rubella at Joint Base San Antonio-Lackland? Of course not, because there were none.

Athletes' measles vaccine rates didn't hit the magic number, but no "epidemics."

There was also a study that looked at measles immunity among players in Major League Baseball (MLB) and the National Basketball Association (NBA). Twelve percent of MLB players did not have "adequate immunity" and 8 percent of NBA players tested did not have adequate immunity. It was even worse for mumps and rubella. Have you heard about measles, mumps, or rubella outbreaks amongst these athletes? No, because they don't exist.

Here's another study that looked at seventy-seven patients with Crohn's disease, one with indeterminate colitis, and forty-five with ulcerative colitis. According to the study, "Sixteen (13.1 %) patients lacked detectable immunity to measles, and four (3 %) had equivocal immunity." Where are the measles outbreaks among this population?

The bottom-line is that we are told that herd immunity requires 92 percent to 95 percent of people to be vaccinated so that the disease doesn't spread. This simply isn't true because if you look at the entire population (not just two-year-olds) every disease has a vaccination rate well below 90 percent. So we should have epidemics of every disease everywhere. But, we don't.

Sources:

Cleveland, Noa K., Dylan Rodriquez, Alana Wichman, Isabella Pan, Gil Y. Melmed, and David T. Rubin. "Many Inflammatory Bowel Disease Patients Are Not Immune to Measles or Pertussis." *Digestive Diseases and Sciences* 61, no. 10 (October 2016), 2972–2976.

Conway, Justin J., Brett G. Toresdahl, Daphne I. Ling, Nicole T. Boniquit, Lisa R. Callahan, and James J. Kinderknecht. "Prevalence of Inadequate Immunity to

Measles, Mumps, Rubella, and Varicella in MLB and NBA Athletes." *Sports Health* 10, no. 5 (September/October 2018), 406–411.

Hutchins, S. S. "National Serologic Survey of Measles Immunity Among Persons 6 Years of Age or Older." *Medscape General Medicine*, January 2001.

Lewis, Paul E., Daniel G. Burnett, Amy A. Costello, Cara H. Olsen, Juste N. Tchandja, and Bryant J. Webber. "Measles, Mumps, and Rubella Titers in Air Force Recruits." *American Journal of Preventive Medicine* 49, no. 5 (November 2015), 757–760.

Mossong, J., D. J. Nokes, W. J. Edmunds, M. J. Cox, S. Ratnam, and C. P. Muller. "Modeling the Impact of Subclinical Measles Transmission in Vaccinated Populations with Waning Immunity." *American Journal of Epidemiology* 150, no. 11 (December 1999), 1238–1249.

"Typhoid Fever Surveillance and Vaccine Use — South-East Asia and Western Pacific Regions, 2009–2013." Centers for Disease Control and Prevention. Last modified October 3, 2014. https://www.cdc.gov/mmwr/preview/mmwrhtml/mm6339a2.htm.

SECRET #33

The Rates of Contagious Diseases Have Come Down Because of Better Public Health

Quick Version: The rates of all of the contagious diseases came down substantially long before vaccines existed because better sanitation made the difference, not vaccines.

The US population at large has immunity rates far below 95 percent for every communicable disease tested. But rates of these diseases have still come down. How is that possible?

Deaths from contagious diseases decreased substantially before vaccines existed.

Here's a chart using US Government data that shows that deaths from measles and other diseases were declining rapidly before the vaccine. How did that happen?

What changed in the twentieth century in the United States that could account for the decrease in all these diseases before vaccines existed? The answer is better hygiene, clean water, and better sanitation.

Consider this—both scarlet fever and typhoid disappeared without a vaccine.

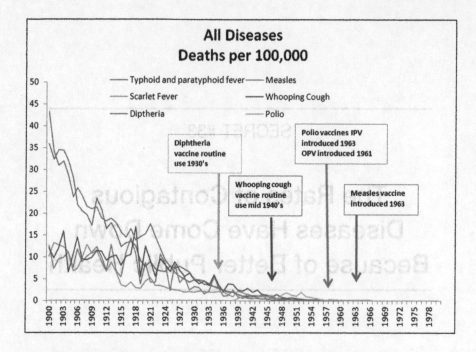

Let's think about typhoid.

Typhoid fever is most common in non-industrialized countries in Asia, Africa, Eastern Europe, and Latin America. According to the CDC:

> Typhoid fever is caused by the bacterium Salmonella enterica serovar Typhi (Typhi). Infection is transmitted via the fecal-oral route with most cases and deaths occurring among populations that lack access to safe drinking water and adequate sanitation and hygiene.

In other words, clean water (that doesn't have human waste) is what is needed to stop typhoid. We developed clean water supplies in the United States in the twentieth century and typhoid disappeared without vaccination. But, typhoid still exists in countries with poor sanitation.

Measles and scarlet fever were down by 81 percent in only nineteen years before vaccines.

Here's an interesting article that appeared in the *Deseret News* of February 27, 1931. The article is titled "America Sets Up New Health Record in 1930." The article states:

> The communicable diseases of childhood – measles, scarlet fever, whooping cough and diphtheria declined remarkably in 1930. Not only did the combined mortality from these four diseases drop 26% in a single year, but each of the four registered a new low death rate. Comparison with the year 1911 shows a 79% decline in the combined mortality of the group, a decline of 81% for measles and for scarlet fever, of 73% for whooping cough, and of 79% for diphtheria.

So, there was an 81 percent decline in the rates of measles and scarlet fever over nineteen years *before vaccines*. In addition, scarlet fever has disappeared in the United States. It is not even a reportable disease. Has there ever been a scarlet fever vaccine? No.

Vaccination is given the credit for eliminating diseases that were on the decline in the United States long before the vaccines were even invented. Herd immunity is not what is critical to reduce the rates of these diseases. What is critical is good hygiene and nutrition, excellent sanitation, and clean water. These are the changes that brought these disease rates down before vaccines existed.

These are also the changes that are needed today in poorer countries. The Bill and Melinda Gates Foundation committed ten billion dollars over ten years to help research, develop, and deliver vaccines for the world's poorest countries. Why not spend that money on clean water, improved sanitation, better nutrition, and helping these countries raise their standard of living? These are the changes that reduced contagious diseases in the United States. The same thing would work in the world's poorest countries. Strangely, the Gates Foundation is chiefly focused on vaccines. Is that because of their financial partnerships with all the big vaccine makers?

Sources:

"CDC: With Low Vaccine Rates, Some Areas Risk Losing Herd Immunity." Advisory Board. Last modified October 21, 2014. https://www.advisory.com/daily-briefing/2014/10/21/cdc-with-low-vaccine-rates-some-areas-risk-losing-herd-immunity.

Deseret News (Salt Lake City, UT). "America Sets Up New Health Record in 1930." February 27, 1930.

Guyer, B., M. A. Freedman, D. M. Strobino, and E. J. Sondik. "Annual Summary of Vital Statistics: Trends in the Health of Americans During the 20th Century." *Pediatrics* 106, no. 6 (2000), 1307–1317.

Lewis, Paul E., and Daniel Bernett. "Measles, Mumps, and Rubella Titers in Air Force Recruits: Below Herd Immunity Thresholds?" *PubMed*. Last modified July 7, 2015.

"Measles: Answers to Common Questions." Centers for Disease Control and Prevention. Last modified August 23, 2019. https://www.cdc.gov/measles/about/faqs.html.

Meissner, H. C. "Why is Herd Immunity So Important." *American Academy of Pediatrics*. Last modified May 2015. DOI: https://doi.org/10.1542/aapnews.2015365–14.

SECRET #34

The Measles Vaccine Can Actually Cause Measles

> *Quick Version:* Vaccines can cause the very disease they are designed to prevent. The person who got vaccinated can get the disease from the vaccine. And the people who come in contact with the vaccine recipient can also get the disease through "shedding."

Another reason why shaming or forcing people into vaccination is wrong is because the measles vaccine can cause measles. That's right, the vaccine itself can cause the very disease it's designed to prevent.

Remember what a vaccination is. It's a small infection of that same disease that you're trying to prevent. In the case of measles, it's a live virus. So it shouldn't be a surprise that a small live infection of measles can sometimes cause . . . measles!

Even though the virus is attenuated (weakened), it can, sometimes, cause the actual disease.

In fact, there are two ways a vaccine can cause the disease it was designed to prevent:

1. The virus in the vaccine causes the disease in the person who got the vaccine.

2. The virus in the vaccine is shed and causes the disease in other
   people.

Shedding happens when the live virus a person gets from an infection or a
vaccine is spread to other people.

For example, here's what scientists from the Department of Medicine, at
the University of California, San Francisco wrote: "Because live, attenuated
viruses are shed from vaccinees, they sometimes present a risk to unvaccinated individuals with impaired immunity."

In other words, kids who are sick can get measles and other diseases like
mumps, rubella, vaccinia, varicella, zoster, yellow fever, rotavirus, and flu
from children who were recently vaccinated. Why? Because these vaccines
contain live viruses and these viruses can be "shed."

The next time you hear about how everyone needs to be vaccinated to
protect the children who are too sick or too young to be vaccinated, think
about this—vaccinations can also spread the disease, especially to vulnerable kids.

Here's a study that involved the CDC. It looked at the urine of children
administered the measles vaccine. The study found that 83 percent of the
kids had the measles virus shed in their urine:

- For children fifteen months old, 83 percent (ten out of twelve)
  children had measles virus RNA that was detected in their urine for
  up to fourteen days after measles vaccination. Why fourteen days?
  Because they stopped looking after that.
- Urine specimens were also obtained from four healthy young adults
  (ages twenty-one to thirty-two years) who were recently vaccinated
  with measles booster shots. Measles virus RNA was detected in the
  urine of all four individuals.

Children are not the best at washing their hands after urinating. Measles
virus may be spread, especially when people are sharing the same household
or bathroom facilities.

Examples of other vaccines that shed and can spread the very disease the vaccine was designed to prevent

Here are examples from other live virus vaccines. You can see that many
people who are vaccinated risk spreading the disease to non-vaccinated people through shedding:

Influenza (Flu Vaccine)—The manufacturer tells us that it is common for people who are vaccinated to shed the virus and it's more common among children. The manufacturer states:

> Shedding of the live attenuated vaccine virus is common after receipt of Live Attenuated Influenza Vaccine. In general, shedding is more common among younger recipients, among whom it may also be of longer duration.

Rubella vaccine—Here we are told that excretion of the rubella virus from the nose or throat happens in the majority of cases and the virus is shed in breast milk too. The manufacturer states:

> Excretion of small amounts of the live attenuated rubella virus from the nose or throat has occurred in the majority of susceptible individuals 7 to 28 days after vaccination . . . transmission of the rubella vaccine virus to infants via breast milk has been documented.

Varicella (chicken pox) vaccine—Here we are told that chickenpox shedding is rare, but we are also told people who are vaccinated should avoid people who can get injured by the virus. The people who can be injured from people who were vaccinated include: pregnant women, newborns, and people who are immunocompromised.

Remember, these are the very same people we are told we must protect by getting vaccinated. The truth, however, is that we put them at risk by getting vaccinated. Vaccine proponents can't have it both ways. The manufacturer states that people who get the vaccine should stay away from susceptible people for up to six weeks:

> Post-marketing experience suggests that transmission of vaccine virus may occur rarely between healthy vaccinees who develop a varicella-like rash and healthy susceptible contacts. Transmission of vaccine virus from a mother who did not develop a varicella-like rash to her newborn infant has been reported. Due to the concern for transmission of vaccine virus, vaccine recipients should attempt to avoid whenever possible close association with susceptible high-risk individuals for up to six weeks following vaccination with VARIVAX. Susceptible high-risk individuals include:
> - Immunocompromised individuals;
> - Pregnant women without documented history of varicella or laboratory evidence of prior infection;

- Newborn infants of mothers without documented history of varicella or laboratory evidence of prior infection and all newborn infants born at <28 weeks gestation regardless of maternal varicella immunity.

Rotavirus vaccine—This is another example where people who are vaccinated shed the virus and the virus can injure people, especially those who are immunocompromised. We are told that we must get vaccinated to protect the children who cannot be vaccinated, but we put these vulnerable kids at risk by getting vaccinated ourselves! The manufacturer states:

> Vaccine virus transmission from vaccine recipient to nonvaccinated contacts has been reported. Caution is advised when considering whether to administer RotaTeq to individuals with immunodeficient close contacts such as:
> - Individuals with malignancies or who are otherwise immunocompromised;
> - Individuals with primary immunodeficiency; or
> - Individuals receiving immunosuppressive therapy.

Smallpox (Vaccinia) vaccine—Smallpox can also be shed by people who are vaccinated. In fact, the CDC tells us:

> Unintended transmission of vaccinia virus can occur through contact with civilian and military personnel vaccinated under the U.S. Department of Defense smallpox vaccination program.

Zostavax (Shingles) Vaccine—According to the manufacturer, "Transmission of vaccine virus may occur between vaccinees and susceptible contacts." Vaccinees are the people who are vaccinated. Susceptible contacts are left to us to guess as they are not described.

Examples of children who got measles or a rash from the measles vaccine

Remember there are two ways that a vaccine can cause the disease it was designed to prevent—shedding to other people and actually causing the disease in the person who got the vaccine.

Here are some examples from the government's own database (VAERS) where children actually got measles from the measles vaccine. They didn't get it because it was shed from another child; they got it directly from the vaccine itself.

<u>Boy, Twenty-two months</u>
Diarrhea following vaccine—immediate . . . Low grade fever—climbed to
102 for few days. Very swollen glands. Change in behavior—foggy—lethargic—lack of interest. Not sleeping—(twitching in sleep). Loss of appetite—
(lost more than 1 pound in 4 days). Full body measles rash (4 days) puffy
cheeks/eyes, eyes photosensitive to sun, bright light. 1 month plus. VAERS
Case # 547392

<u>Boy, Thirteen months</u>
Evening after immunization fever of 103 degrees started and persisted x 2
days. Onset of red rash on body 1 week later. Suspect was the measles rash.
VAERS Case #521162

<u>Girl, Fifteen months</u>
Patient had approximately 20 lesions. At first developed bumps at vaccination site then developed rash on trunk, back and face. Some spots are
starting to scab over. No fever, no s/s of cellulitis or pneumonia. VAERS
Case # 521432

<u>Girl, Four years old</u>
Severe skin rash; thought to be something similar to erythema multiforme
major; lasted 2 months, extreme tightening; started on arms and moved to
legs; red circles looked like rings; red and raised. VAERS Case # 520725

Getting vaccinated can also spread disease.

Let's think about this. We are told that we must vaccinate our children
to protect our neighbor or the child who cannot be vaccinated. Does that
really make sense?

First, the vaccine can be shed and put our neighbor at risk. Second,
the vaccine itself can cause the disease it was designed to prevent. A
healthy child gets the measles vaccine and gets measles. Now that child
is theoretically more of a danger to his neighbor than before they were
vaccinated.

The idea of vaccinating everyone with a product that doesn't always
work, can cause the very disease it was designed to prevent, and can be shed
and infect other people doesn't make sense because it doesn't protect your
neighbor. In fact, it can put your neighbor at greater risk.

Sources:

"Current Circular Showing Revisions for Porcine Circovirus Removal of Statements Regarding PCV-1 and PCV-2 DNA." U.S. Food and Drug Administration. Accessed February 12, 2021. https://www.fda.gov/media/75718/download.

"Highlights of Prescribing Information M.M.R. II." U.S. Food and Drug Administration. Accessed February 12, 2021. https://www.fda.gov/media/75191/download.

"Highlights of Prescribing Information—Rota Teq." *Procon.org.* Accessed February 15, 2021. https://images.procon.org/wp-content/uploads/sites/17/rotateq-package-insert-2013.pdf.

"Highlights of Prescribing Information—VARIVAX." U.S. Food and Drug Administration. Accessed February 12, 2021. https://www.fda.gov/media/76000/download.

Lauring, Adam S., Jeremy O. Jones, and Raul Andino. "Rationalizing the development of live attenuated virus vaccines." *Nature Biotechnology* 28, no. 6 (June 2010), 573–579.

Rota, P. A., A. S. Khan, E. Durigon, T. Yuran, Y. S. Villamarzo, and W. J. Bellini. "Detection of Measles Virus RNA in Urine Specimens from Vaccine Recipients." *Journal of Clinical Microbiology* 33, no. 9 (September 1995), 2485–2488.

"Safety of Influenza Vaccines." Centers for Disease Control and Prevention. Last modified November 7, 2019. https://www.cdc.gov/flu/professionals/acip/safety-vaccines.htm.

"Secondary and Tertiary Transmission of Vaccinia Virus After Sexual Contact with a Smallpox Vaccinee—San Diego, CA." Centers for Disease Control and Prevention. Last modified March 1, 2013. https://www.cdc.gov/mmwr/preview/mmwrhtml/mm6208a2.htm.

One-Size-Fits-All Vaccine Programs Do Not Work and Are Unsafe for Children Because (Surprise) All Children Are Not All the Same

> *Quick Version:* The one-size-fits-all mass vaccine campaigns forced on us by the government are dangerous because children are not all biologically or genetically identical, and they react differently to the same vaccine.

Have you ever known anyone who got a prescription drug from their doctor, took the medicine, and got a side effect?

Chances are you know plenty. Maybe even yourself? Maybe your child?

Does that mean everyone gets the same side effect from every drug? No. Some people do and some people don't.

Have you ever known somebody who got a drug from their doctor and it didn't work? You probably know plenty.

It's the same with vaccines. Some children will get short-term or long-term side effects and some will not. For some, the vaccine will work, and for some it will not.

Why is that?

It's simple. People are not biologically and genetically identical.

Children are not genetically identical.

If you have more than one child you know this. Your children are living together under the same roof. One gets the flu, and one doesn't. One gets sick with X, and one gets sick with Y.

Dr. Sarah Kim-Hellmuth is a scientist at the New York Genome Center, Columbia University and the Max Planck Institute in Munich. She studies how human genes interact with people's immune systems. She and her colleagues have said: "The human immune system plays a central role in autoimmune and inflammatory diseases, cancer, metabolism, and aging."

This means that diseases such as cancer, inflammation, and autoimmune diseases are controlled, to a large extent, by the individual's immune system. That's not surprising, but here is where it gets interesting. She and her colleagues stated that: "These genes include many of the well-known genes of the human immune system, demonstrating that genetic variation has an important role in how the human immune system works."

This means that how your immune system reacts to viruses or bacteria depends on your genes! (Remember, viruses and bacteria are in vaccines.)

Dr. Kim-Hellmuth states, "These genes include many of the well-known genes of the human immune system, demonstrating that genetic variation has an important role in how the human immune system works."

Dr. Kim-Hellmuth's colleague is Dr. Tuuli Lappalainen from the New York Genome Center and Columbia University. Dr. Lappalainen stated, "This supports a paradigm where genetic disease risk is sometimes driven . . . by causing a failure to respond properly to environmental conditions such as infection."

What does all this mean?

It means that "hundreds of genes" control how an individual's immune system works, and the immune system plays a central role in determining how an individual reacts to an infection.

Vaccination is a type of infection.

Therefore, how children react to a vaccine depends on their individual genes.

Some children will get sick and others will not. Some will get the disease and others not.

Since there are hundreds of genes involved in the immune system, the variability in how individual children deal with infections from vaccines is immense.

The government wants us to believe that a one-size-fits-all mass vaccination campaign is safe for everyone. That simply is wrong because it ignores the science of genetic individuality. We all have hundreds of genes that affect how our individual immune systems work and how we react to a vaccine depends on our individual immune system.

One child may deal with a vaccination (i.e., infection) perfectly fine, and another child may become terribly sick. It all depends on their genetic make-ups that control their immune systems.

Vaccine makers and the government ignore children's genetic individuality.

We are all individuals. Children are genetically different from one another. How an individual reacts to an infection (i.e., a vaccine) depends on their immune system and their immune system depends on their genes. Everyone's genes are different.

The government and their vaccine proponents are ignoring this basic truth. Even when they admitted that some children are prone to mitochondrial disease and this can lead to autism, they decided not to research ways of identifying these children.

It's much more convenient for the organizations and people making millions or billions of dollars from vaccines to ignore the science that says genetics affect our immune systems. They really are not interested in protecting children's health. If they were, they would be spending money and time learning how to vaccinate safely by taking into consideration the fact that children have genetic variability which directly affects how they respond to vaccination. But, they don't do that.

Sources:

Cohen, Elizabeth. "This Mom Wants You to Know What Measles Did to Her Baby." CNN. Last modified May 6, 2019. https://www.cnn.com/2019/05/06/health/measles-baby-misdiagnosis-eprise/index.html.

"Genetic Variants Found to Play Key Role in Human Immune System: Genetic Differences in Immune Response Demonstrate Interaction of Genetics and Environment Linked to Disease Risk." ScienceDaily. Last modified February 11, 2021. https://www.sciencedaily.com/releases/2017/08/170816084913.htm.

"Genetics Play Major Role in Infectious Disease Susceptibility." Healio. Accessed February 12, 2021. https://www.healio.com/news/infectious-disease/20120225/genetics-play-major-role-in-infectious-disease-susceptibility.

Kim-Hellmuth, S. "Genetic Regulatory Effects Modified by Immune Activation Contribute to Autoimmune Disease Associations." Nature. Last modified August 16, 2017. https://www.nature.com/articles/s41467-017-00366-1.

CONCLUSION

You Should Not Be Forced to Vaccinate

Quick Version: What vaccines really do and what the government says they do are two totally different things.

A vaccine like measles is designed to prevent a rash. The vaccine may or may not work and may or may not cause small or large side effects; therefore, shouldn't you have the right to decide whether to have it administered to your child?

As parents, shouldn't we be allowed to weigh the pros and cons of the vaccine and make our own decision?

Or should all children be forced to get all seventy-plus vaccines according to a one-size-fits-all schedule? That's what the government and the drug companies want. Each child is a profit center for Big Pharma and when children get injured, the drug company doesn't pay, we do.

Isn't it obvious by now why Big Pharma, the government, and the media use fear to push vaccines?

They are denying the connection to side effects.

They are denying the fact that children are genetically different from one another.

They are telling scientists not to investigate susceptible groups.

2220202202 202 202 202 202 202 202 202 202 202 202 202 202   The Measles Book

They are telling us that we have to vaccinate our children in order to protect our vulnerable neighbor even though we put our neighbor at risk with shedding.

They tell us vaccination protects us even though it may not work and can cause the very disease it was designed to protect against.

They tell us that measles is an epidemic and, after laws are changed removing exemptions, we are told that measles is still eradicated.

We are told that there is no connection between vaccines and auto-immune diseases even though scientific organizations like the Institute of Medicine published that there were connections twenty-nine years ago.

We are told vaccines are safe even though the government has already paid out nearly four billion dollars to vaccine-injured people.

We are told there is no connection between vaccines and autism even though the science shows how the measles vaccine can destroy the myelin around nerve cells. The government has compensated a child who got autism from vaccines, and has compensated other vaccine injured children who labeled their autism "encephalitis."

We are told that vaccines eliminated all kinds of diseases even though most of these diseases had almost disappeared long before vaccines were invented, and some of them disappeared entirely without vaccines.

We are told vaccines are "unavoidably unsafe" even though manufac-turers have made no effort to make their older vaccines any safer even after more than half a century.

We are told to trust the vaccine manufacturers even though they are not held accountable when a child is hurt or killed by one of their vaccines, and these companies have acted illegally and criminally over many years.

We are told to be afraid of measles even though it's basically a rash.

We are told to ignore the fact that 250,000 people die every year in the United States from medical errors.

We are told to trust our health authorities even though they get paid and in some individual cases get rich from vaccines.

We are told vaccines are thoroughly tested for safety even though they use scientific dishonesty to cover up dangerous side effects by not using a real placebo.

We are told that the scientific establishment is 100 percent behind the safety and effectiveness of vaccines when there are many thousands of sci-entific and medical articles that discuss injuries and deaths from vaccines.

This misinformation that is spoon-fed to us and the fear that is created and disseminated by the media distracts us from making the best decision for our families.

The media, the government, and the pharmaceutical companies want you to be scared. They use fear tactics and lies to scare us into submission. When that doesn't work, they change the laws to force us to submit.

When you hear some politician or doctor make blanket statements like:

"There is no link to autism."

"Vaccines are completely safe."

"Vaccines are our greatest public health success story."

"If you don't vaccinate, your child may die or get cancer."

Ask yourself—are they being honest? Why are they ignoring all the evidence to the contrary and the hundreds of thousands of injured children?

What's in it for them?

Most pediatricians are well meaning, but nearly all the information they have received about vaccines come from vaccine manufacturers and the government. They probably don't know any of the information that's in this book. Most don't even know what's in the vaccine. When asked, they admit that they have no idea of the individual ingredients in the vaccine.

Other vaccine propagandists are not so innocent. Some of the people who are advocating for vaccines make their living from getting people to vaccinate by fear or force. If vaccines didn't exist or if vaccine rates plummeted, they would have to find another way to make money.

We urge you to check the sources in this book. Read them for yourself. Do your own research. Take control of your own child's health; don't blindly hand over the reins. Remember that the government, the media, and Big Pharma are focused on profits. Never let any company use your child as a profit center. Your child's health is your responsibility, not theirs.

What You Can Do to Protect Your Child and Your Rights

> *Quick Version:* You can protect your children's health and your rights as a parent if you stand up with the tens of thousands of other parents who are saying "Enough is enough! You don't own our children's bodies!"

Today, the vaccine makers and their allies in government and the media are forcing our children to undertake an immense medical experiment for their benefit, not ours. Vaccine exemptions are being taken away in various states. However, the tide is turning. People have woken up.

There are many laws in the United States and internationally that state that people should not be subject to experimental medical procedures against their will.

So what can you do?

- Educate yourself before you vaccinate. Look at the resources at Children's Health Defense, NVIC, and other reputable educational websites.
- Read the insert. This is the information sheet that comes with every vaccine. You can find them online or ask your doctor.
- Research the VAERS database. Become familiar with the injuries and deaths associated with different vaccines.

Learn the facts about vaccines, get active, and join groups like:

- Children's Health Defense: https://childrenshealthdefense.org/
- National Vaccine Information Center: https://www.nvic.org/
- Informed Consent Action Network (ICAN): https://www .icandecide.org/
- Stand for Health Freedom: https://standforhealthfreedom.com/

Take a copy of this book and give it to your state's representatives.

Remember that the health of your children is your responsibility. Do not allow politicians and companies with financial interests to make medical decisions for them!

Glossary

Acute disseminated encephalomyelitis (ADEM): Inflammation of the brain and spinal cord which damages the covering of nerve fibers as an autoimmune response of the body.

Acute hemorrhagic edema of infancy (AHEI): Fever, a bruise-like rash, and swelling on the face, ears, and limbs that affects children younger than two years.

Afebrile convulsions or seizures: Sudden, uncontrolled disturbance in the brain without the presence of a fever.

Anaphylactic shock: A severe whole-body allergic reaction that causes a sudden drop in blood pressure and narrowing of the airways that can be life-threatening. Poor circulation and a lack of oxygen and nutrients cause the body to go into a state of shock.

Anaphylaxis: A severe whole-body allergic reaction that is potentially life-threatening.

Anaphylaxis and anaphylactoid reactions: Hypersensitivity reactions are anaphylactic if immune mediated and anaphylactoid if chemically mediated. These reactions are produced through different pathways in the body but present with the same symptoms.

Angioneurotic edema: A genetic form of swelling below the skin layers.

Arthralgia: Pain in a joint.

Arthritis: Painful swelling and stiffness of the joints.

Aseptic meningitis: Inflammation of the meninges, the protective covering of the brain and spinal cord that is not caused by a bacterial infection.

Asystole: The state where the heart is not producing electrical activity that results in no heartbeat or blood flow throughout the body. Shown as a flat line on an EKG and is usually irreversible.

Ataxia: Lack of muscle control, coordination, or balance.

Attenuated: A weakened form.

Atypical measles: A form of measles that presents after immunization of a killed measles vaccine and then exposed to wild-type measles.

Brachial neuritis: Nerve damage that causes pain, weakness, or loss of function in muscles of the chest, shoulder, arm, or hand.

Bradycardia: Heart rate that is lower than normal.

Bronchial spasm: Sudden muscle tightening of the small airways in the lungs.

Cardiopulmonary resuscitation: Chest compressions given to a person in an attempt to restore blood flow to the heart and breathing.

Chronic fatigue syndrome: Severe tiredness that cannot be explained.

Chronic inflammatory demyelinating polyneuropathy: A disorder that damages the protective covering of nerves, causing weakness and loss of sensation in the legs and arms.

Conjunctivitis: Swelling of the thin layer of tissue on the inner surface of the eyelid.

Cough: A sudden and forceful release of air from the airway and mouth.

Cuffed endotracheal (ET) tube: A tube is placed within the nose or mouth and into the trachea to create an airway and restore breathing. The balloon at one end of the ET tube can be inflated for proper placement and creates a cuff to prevent contents from entering or exiting.

Cyanosis: A blue discoloration of the skin from a lack of oxygen or poor blood circulation.

Cyanotic: The state of having a blue discoloration of the skin from a lack of oxygen or poor blood circulation.

Diabetes mellitus: A disorder of high blood sugar over an extended time because the body is unable to produce enough insulin.

Diarrhea: Loose, watery bowel movements.

Diffuse retinopathy: Disease of the retina and surrounding structures or tissues that result in vision impairment or loss of vision.

Dizziness: Abnormal sense of balance and place.

Erectile dysfunction (ED): Inability to get and keep an erection firm enough for sexual intercourse.

Encephalitis: Swelling of the brain.

Encephalopathy: Any brain disease, damage, or malfunction.

Endogenous avian leukosis viruses: A virus that can cause various diseases, tumor formation, and death in chickens.

Epididymitis: Swelling and irritation of the tube on the back of the testicle that stores and carries sperm.

Epinephrine: Medication used to treat severe allergic reactions and to restore heart and breathing functions during emergencies.

Erythema: Abnormal redness of the skin.

Erythema multiforme: An allergic reaction that causes a "target" shaped skin condition.

Erythema multiforme major: A severe and potentially life-threatening allergic reaction that causes a red and raised "target" on the skin all over the body. Also known as Stevens-Johnson syndrome.

Etiologic: The cause or factor to the development of a disease or condition.

Febrile: Having or showing symptoms of a fever.

Febrile seizures (febrile convulsions): A convulsion due to a spike in body temperature.

Fever: A body temperature higher than normal.

Fibromyalgia: Pain and weakness of the muscles and bones that often occurs with extreme tiredness, sleep, memory, and mood disturbances.

Flare: Pain and inflammation at the site of medication injection.

Gait disturbances: Abnormal pattern of walking.

Gianotti-Crosti syndrome (GCS): A childhood skin condition with a red raised rash and blisters on the skin of the cheeks, buttocks, arms, and legs that results from a response to a previous viral infection.

Guillain-Barré Syndrome (GBS): The body's immune system attacks the nerves that are outside the brain and spinal cord, causing weakness, tingling, numbness, and eventually paralysis.

Hemic and lymphatic system: Structures, such as the spleen, bone marrow, and stem cells that are involved in the production and filtration of blood and roles of the immune system.

Hemophagocytic lymphohistiocytosis: A cancer-like disorder where the body's immune system attacks tissues or organs and can be life-threatening.

Henoch-Schönlein purpura: A disease involving inflammation of small blood vessels in the skin, intestines, kidneys, and joints to leak and form a rash.

Hepatitis: Inflammation of the liver.

Home Health Nurse (HH Nurse): A nurse who provides care to people in their home.

Induration: Localized hardening of soft tissue.

Intubation: The insertion of a tube into a person's body.

Intraosseous infusion (IO): The process of injecting directly into the marrow of a bone.

Irritability: An abnormal or excessive response of an organ or body part.

Leukocytosis: An increase of white blood cells in the blood during infection.

Malaise: A general feeling of discomfort or illness with an unknown cause.

Measles inclusion body encephalitis: A disease that causes inflammation of the brain and occurs in a weakened immune system state and after measles infection or vaccination given.

Measles-like rash: A rash of small red spots, some of which are raised, often seen in a measles infection.

Meningitis: Inflammation of the protective coating of the brain and spinal cord.

Multiple sclerosis: Damage to the protective covering of nerve cells in the brain and spinal cord that delays or blocks messages between the brain and the body.

Myalgia: Muscle pain.

Nerve deafness: Loss of hearing or impairment from nerve damage in the inner ear.

Neuromyelitis optica: A disorder where the body's immune system attacks eye nerves and the spinal cord.

Neonatal Intensive Care Unit (NICU): A specialized department in the hospital that provides critical care to infants or premature babies.

Ocular palsies: Eye movement disorder caused by cranial nerve damage.

Optic neuritis: Inflammation that damages the nerve fibers bundle that transmits visual information from the eye to the brain, causing pain and temporary vision loss.

Orchitis: Inflammation of one or both of the testicles.

Otitis media: Inflammation of the middle ear.

Pancreatitis: Inflammation of the pancreas.

Panniculitis: Inflammation of the fatty layer under the skin.

Papillitis: Inflammation and deterioration of the portion of the nerve that enters the retina of the eye.

Paresthesia: Abnormal sensation of the body that feels like tingling, numbness, or pricking.

Parotitis: Inflammation of the parotid gland, a gland that produces saliva.

Pneumonia: Inflammation of the lungs air sacs caused by a bacterial or viral infection.

Pneumonitis: Inflammation of the walls of the air sacs in the lungs.

Polyneuritis: Any disorder that affects the nerves that run through the body and outside the brain and spinal cord.

Polyneuropathy: Damage to any nerves that run through the body and outside the brain and spinal cord.

Postvaccinal parkinsonism: Signs and symptoms identical to diagnostic standards of Parkinson's disease that occur after receiving a vaccine.

Pruritus: Severe itching of the skin.

Rash: Change of color, appearance, or texture of the skin.

Recombinant human albumin: A genetically engineered product that is structurally equivalent to human serum albumin.

Regional lymphadenopathy: Lymph nodes of abnormal size or consistency involving a single node or multiple nodal regions.

Retinitis: Inflammation of the retina, the back layer of the eyeball that regulates the amount of light allowed into the eye, and the formation of images.

Retrobulbar neuritis: An inflammation of the optic nerve behind the eyeball.

Reye's syndrome: A condition that causes swelling in the liver and the brain after a viral infection that mostly affects children and teenagers.

Rhinitis: Inflammation of the mucous membrane of the nose.

Sensorineural hearing loss: A type of hearing loss caused by damage to the inner ear, sensory organs, or the associated nerve.

Sequelae: A condition as a result of a previous disease or injury.

Somnolence: The state of sleepiness, feeling drowsy, or a readiness to fall asleep.

Signs and symptoms (S/S): Any abnormality showing a medical condition.

Stevens-Johnson syndrome: A severe whole-body allergic reaction to a medication that creates painful blisters and lesions on the skin and mucous membranes and severe eye problems.

Subacute sclerosing panencephalitis (SSPE): A chronic brain disease of children and young adults occurring after an attack of measles, causing seizures, mental retardation, abnormal movement, and often death.

Swelling: Abnormal enlargement of a body part, usually as a result of fluid accumulation.

Syncope: Partial or complete loss of consciousness that is known as fainting.

Tenderness: Sensitivity to pain.

Thrombocytopenic purpura: A condition that causes platelets, the blood cells that cause clotting, to become destroyed by the immune system.

Toxic shock syndrome: A life-threatening condition caused by a bacterial infection of menstruating women who use tampons.

Transverse myelitis: The inflammation of both sides of one section of the spinal cord.

Urogenital system: An organ system of the reproductive organs and the urinary system.

Urticaria: A rash of round, red, itchy welts on the skin caused by an allergic reaction.

Vagal: Relating to the vagus nerve, a cranial nerve that supplies to organs of the chest and abdomen.

Vasculitis: Inflammation of one or more blood vessels.

Vesiculation: The formation of blisters.

Wheal: A lump at the site of injection before the solution is absorbed.